SUSTAINABLE NATURAL RESOURCE MANAGEMENT FOR SCIENTISTS AND ENGINEERS

Natural resources support all human productivity. Their sustainable management is among the preeminent problems of the current century. Sustainability and the implied professional responsibility start here. This book uses applied mathematics familiar to undergraduate engineers and scientists to examine natural resource management and its role in framing sustainability. Renewable and nonrenewable resources are covered, along with living and sterile resources. Examples and applications are drawn from petroleum, fisheries, and water resources. Each chapter contains problems illustrating the material. Simple programs in commonly available packages (Excel, MATLAB) support the text and are available for download from the Cambridge University Press website. The material is a natural prelude to more advanced study in ecology, conservation, and population dynamics, as well as engineering and science. The mathematical description is kept within what an undergraduate student in the sciences or engineering would normally be expected to master for natural systems. The purpose is to allow students to confront natural resource problems early in their preparation.

Daniel R. Lynch is the MacLean Professor of Engineering Sciences at Dartmouth College and Adjunct Scientist at the Woods Hole Oceanographic Institution. Through the 1990s he served on the Executive Committee of the US GLOBEC Northwest Atlantic Program and cofounded the Gordon Research Conference in Coastal Ocean Modeling. He has published extensively on finite element methods in coastal oceanography and is coeditor of the AGU volume *Quantitative Skill Assessment for Coastal Ocean Models* and a related volume, *Skill Assessment for Coupled Physical-Biological Models of Marine Systems*, published as a special volume of the *Journal of Marine Systems*. In 2004 he wrote a graduate textbook titled *Numerical Solution of Partial Differential Equations for Environmental Scientists and Engineers: A First Practical Course*. At Dartmouth's Thayer School, Dr. Lynch developed the Numerical Methods Laboratory around the theme of interdisciplinary computational engineering. He pursues research at the intersection of advanced computation and large-scale environmental simulation. Current investigations focus on sustainability, natural resources, and professional education.

SUSTAINABLE NATURAL RESOURCE MANAGEMENT FOR SCIENTISTS AND ENGINEERS

Daniel R. Lynch

Dartmouth College

CAMBRIDGE UNIVERSITY PRESS
Cambridge, New York, Melbourne, Madrid, Cape Town, Singapore, São Paulo, Delhi

Cambridge University Press
32 Avenue of the Americas, New York, NY 10013-2473, USA

www.cambridge.org
Information on this title: www.cambridge.org/9780521899727

First published 2009

Printed in the United States of America

A catalog record for this publication is available from the British Library.

Library of Congress Cataloging in Publication data
Lynch, Daniel R.
   Sustainable natural resource management for scientists and engineers / Daniel R. Lynch.
     p. cm.
   Includes bibliographical references and index.
   ISBN 978-0-521-89972-7 (hbk.)
   1. Natural resources–Management–Mathematical models.   2. Natural resources–Environmental
aspects–Management.   I. Title.
   HC85.L96 2009
   333.7–dc22   2008053176

ISBN 978-0-521-89972-7 hardback

CONTENTS

TO BEGIN

... And God saw that it was good. Then God said, "Let us make man in our image, after our own likeness; and let them have dominion over the fish of the sea, and over the birds of the air, and over the cattle, and over all the earth, and over every creeping thing that creeps upon the earth." So God created man in his own image, in the image of God he created him; male and female he created them. And God blessed them, and God said to them, "Be fruitful and multiply, and fill the earth and subdue it; and have dominion over the fish of the sea and over the birds of the air and over every other living thing that moves upon the earth. ... And God saw everything that he had made, and behold, it was very good.

> Genesis 1: 25–28, 31. The Bible, Revised Standard Version

I rode through the "Schroon Country" with a man who has probably done as much as anyone to desolate this whole region ... As league after league of utter desolation unrolled before and around us, we became more and more silent. At last my companion exclaimed: "This whole country's gone to the devil, hasn't it?" I asked what was, more than anything else, the reason or cause of it. After long thought he replied: "It all comes to this – it was because there was nobody to think about it, or to do anything about it. We were all busy, and all somewhat to blame perhaps. But it was a large matter, and needed the co-operation of many men, and there was no opening, no place to begin a new order of things here. I could do nothing alone, and my neighbor could do nothing alone, and there was nobody to set us to work together on a plan to have things better; nobody to represent the common object."

> J. B. Harrison, *Garden and Forest 2:74*, July 24, 1889, p 359

Mr. Baker: "As I have talked with thousands of Tennesseeans, I have found that the kind of natural environment we bequeath to our children and grand-children is of paramount importance. If we cannot swim in our lakes and rivers, if we cannot breathe the air God has given us, what other comforts can life offer us?"

Mr. Muskie: "... Can we afford clean water? Can we afford rivers and lakes and streams and oceans which continue to make life possible on this planet? Can we afford life itself? ... These questions answer themselves. ... Let us close ranks ... so that we can leave to our children rivers and lakes and streams that are at least as clean as we found them, and so that we can begin to repay the debt we owe to the water that has sustained our Nation."

Senators Howard Baker and Edmund Muskie, *Congressional Record*, October 17, 1972

... and He walks in His garden, in the cool of the day

"Now is the Cool of the Day," Jean Ritchie, *A Celebration of Life*, 1971

PREFACE

Natural resources support all human productivity; their sustainable management is among the preeminent problems of the current century. Sustainability, and the implied professional responsibility, starts here.

The primary audiences for this book are scientists and engineers. They are among the people whose professional work directly engages natural resources, whether through harvesting, conversion, or conservation. Constructing a sustainable relationship between natural resources and the human activity they support is a problem that must be embraced by this group of professionals. Accordingly, we use their language – intrinsically scientific and mathematical. And we emphasize quantification and analysis as first principles.

The overall objective of this book is to bring together a unified presentation of natural resources. There are three generic elements:

- Dynamics of the resource in question
- Value of the resource and its uses
- Ownership and "control" of outcomes

or loosely in terms of disciplines: natural science, economics, and political science. Each of these must be blended in any resource analysis. They are the framework of sustainability.

There have been many approaches to this general problem, offering important theories and insights from individual disciplinary perspectives. Among them are harvesting, population structure and dynamics, ecology, land use and geography, economics, water, development, agriculture, forestry, and conservation. Each tradition speaks to a different audience and addresses distinct, specific resource issues, utilizing linear algebra, differential and difference equations, optimization, and computation as needed. The varied use of these analytical tools has been conditioned by the audience and the disciplinary setting. But all of them venture into some similar and overlapping territory in describing key resource concepts (harvest, effort, extraction, extinction, consumption, etc.). It is a goal of this text to present these ideas in analytical frameworks natural to science and engineering and, by so doing, to engage these professional groups broadly in the problem of sustainable resource management.

Natural Resource Classification. The book is structured within a simple two-way classification, as illustrated in the table below. In the exhaustible category, the basic descriptor is the *amount S* of the resource; in the renewable category, the *rate of occurrence Q* is paramount. "Sterile" indicates biochemical inactivity, whereas "living" implies self-reproduction. The various quadrants shown are the intersections of these two binary categories:

	Exhaustible S	Renewable Q
Sterile	1 (Oil)	2 (Water)
Living	4 (Fish) ←	→ 3 (Fish)

    Chapter 1 explores the exhaustible sterile category, Quadrant 1, using the example of petroleum throughout. Such a resource will be exhausted eventually; its trajectory is described in terms of the amount available over time. Fundamentally, the resource is finite; some may be undiscovered, but that does not change the facts, only our limited knowledge of them. Discovery is treated as an economic activity, and "Hubbert's Peak" is found in the intersection of utilization, discovery, and demand expansion. Many fundamental concepts of resource economics are exposed in this quadrant, where the finite supply S is the paramount concern. The only sense of sustainability in this quadrant is that associated with the substitute – money invested or knowledge gained – and the legitimacy of the trade-off implied.

    Chapters 2, 3, and 4 add the significant feature of self-reproduction: the living resource. Fisheries are the example. As indicated, this case is a hybrid, occupying both Quadrants 3 and 4. *S produces Q.* It is possible to treat such a resource as exhaustible, "mining" S to extinction. This risk of extinction is ever present with a living resource, enhanced when growth is slow or highly variable and/or the harvesting is unruly. This case also admits many steady, sustainable states, where the stock S is kept constant and the self-renewal rate Q is maintained. Hence, sustainable use of a living resource implies sustaining the conditions that support its continued presence and growth; the harvesting activity must be consistent with that sustained presence.

    The three chapters present increasing biological sophistication, all the while overlapping Quadrants 3 and 4 and sharing this basic duality. Chapter 2 examines the simplest description in terms of a single biomass variable. It exposes features that endure through Chapters 3 and 4, most notably the need to describe the harvesting effort, the technology utilized, and the intersection of economic reasoning on the part of the "owner" and the "harvester." Chapter 3 discusses populations that are structured according to recognizable life stages. Chapter 4 examines the development of cohorts of individuals and concludes with an introduction to individual-based descriptions. A fundamental distinction among these chapters is the way that reproduction is handled. In Chapter 2, it is completely endogenous; Chapter 4 represents it as completely exogenous; and Chapter 3 represents both extremes and helps to put them in context.

Chapter 5 treats the case of the renewable, sterile resource; the standard example is water. In this case, the resource is fundamentally fugitive; S is uncontainable except in very limited amounts and on very short timescales. The rate of occurrence Q is exogenous; we steer Q, but we cannot sequester it for very long.

A third binary axis of classification (not shown in the figure) would be the "degradable/nondegradable" one, where we find the classic case of air or water pollution. **Chapter 6** treats this case in the form of an introduction to pollution and assimilative capacity, building on the networked water description in Chapter 5.

Prerequisites. The present treatment is at the mezzanine (third-year university) level. It requires a first-year university preparation in linear algebra, ordinary differential equations, and computation. Some exposure to operations research and optimization is useful, through linear and mixed-integer programming. That can be introduced here, but ultimately it deserves amplification in separate coursework. Facility with simple computation tools (notably MATLAB and the Excel Solver) is assumed. An exposure to the basic economics of public goods is a valuable supplement. The purpose is to encourage students to confront natural resource problems early in their preparation. At this level, the material has several central themes, which admit a quasi-unified treatment. With this exposure, many in-depth extensions are possible, depending on one's field of interest.

In the terminology of engineering science, the descriptions use lumped system theory. The optimization is in terms of either the steady states of such systems (e.g., regional water resource systems) or their optimal trajectories (e.g., extraction and exploration histories for nonrenewables). The mathematical description is kept within what an undergraduate student in the physical sciences or engineering would normally be expected to master for other natural systems. The focus is on describing dynamic interactions among resources, economies, and ownership agendas that together determine outcomes. Quantitative mastery of model systems and an ability to transfer those dynamics to realistic contemporary problems are the goals.

This material has been offered to undergraduate Science Division students at Dartmouth College. The lectures form the core of a full course in the topic; they should be supplemented with more descriptive readings according to the instructor's design and interest.

The lectures are also useful as a set of supplementary examples for teaching the basic mathematics covered, supporting a "natural resources across the curriculum" deployment. Either mode – a stand-alone course or the diffusion of the material throughout existing curricula – is an essential beginning in the general area of sustainability.

Software. In terms of computation, the text emphasizes current generic platforms: MATLAB and Excel for its optimization package. These are not intended to be prescriptive, but rather to be usable on portable platforms in the lab, office, or boardroom. There are some elementary programs in each chapter offered as examples for these common platforms. These are summarized at the end of each chapter and are available on the publisher's Web site: http://www.cambridge.org/9780521899727

Audiences. There are several audiences for this book:

1. *Undergraduate students of science and engineering.* These lectures originated in this cohort. The material is a critical foundation for understanding and bringing about sustainability. And, as an integrative introduction, it is a gateway to more advanced study in environmental science and engineering, ecology, and population dynamics, and the intersection of these fields with law, economics, business, public policy, and international development.
2. *Graduate students in professional programs* concerned with the development process, technology management, conservation, and the natural resource/economic/societal interaction. For this audience, undergraduate mathematical study is assumed. Computation is liklely to be a most practical and attractive entry point.
3. *Mathematics instructors* in search of lectures treating basic calculus, ODEs, linear algebra, and optimization. The context of natural resources is urgently contemporary. It makes full use of these basic tools and presents a broad front for applied research, particularly in the present context of widespread access to computational power, observational capacity, and networking.

Today, professionals in all walks of life are employing computation on laptops in every setting, with portable programs connecting Web-served databases and information archives to the boardroom. This book projects quantitative natural resource analysis into this arena of common professional activity – a critical step in the implementation of just and sustainable outcomes.

In summary, this text fills the need for a multiresource exposition for undergraduate students of science and engineering. It uses the mathematical preparation already required of these students and introduces many paths into more disciplinary study in ecology, water resources, population dynamics, and resource economics. It is a necessary element in understanding sustainability and the role of science and engineering in achieving it.

Acknowledgments. Many individuals and institutions contributed to this work.

The National Science Foundation supported much of the work represented here, throughout many different projects. That support has been a privilege and has opened many doors.

My colleagues at the Woods Hole Oceanographic Institution, the National Marine Fisheries Service, and the Department of Fisheries and Oceans have been a constant source of inspiration.

Visiting terms were spent at the University of Notre Dame, School of Engineering; the Catholic University of America, School of Philosophy; and Princeton University, Woodrow Wilson School of Public and International Affairs. These were all critical incubation times for the work assembled here, and I am grateful for each of these opportunities.

Finally, words cannot express my appreciation for the support of my immediate family: the bibliophile, the theologian, and the planktoneer. They sustain me, and I dedicate this work to them.

Daniel R. Lynch
Hanover, NH
June 15, 2008

1 Sterile Resources

In this chapter, we introduce the simple conception of a scarce resource, locally owned and globally traded. It is the classic Quadrant 1 resource, valuable and scarce. Production amounts to making it available for sale into an economic market in which its scarcity affects its price. Accordingly, price is endogenous to the resource system. Production is the same as consumption, which is tantamount to destruction: irreversible conversion to other chemical forms with no recycling. The owner's basic decision is how fast to produce.

That the resource is finite is a first principle. The fact that some portion of the resource is undiscovered at any point in time does not change its finiteness. What does change, over time, is the improving state of knowledge about how much of the resource there is. Decisions about how fast to produce are always reached within an environment of imperfect knowledge and speculation about future discoveries. There is a need to make decisions in this uncertain environment and a need to adjust continually as new information becomes available. Exploration reduces, but does not eliminate, uncertainty.

This case is extreme in its simplicity. It is elaborated below; the example of petroleum is used throughout. Many critical concepts of resource economics are introduced and carried forward into subsequent chapters.

1.1 COSTLESS PRODUCTION OF A STERILE RESOURCE

1.1.1 Base Case

This is the simplest case of exhaustion of a finite resource. We will use the terminology

$S(t)$ = amount of the resource remaining to be produced and sold
$X(t)$ = production rate
$P(t)$ = market price per unit of production

It is assumed that the resource is owned unambiguously, that it costs nothing to produce it, and that S, X, and P are known with perfect certainty. Three relations govern:

Mass balance:

$$\frac{dS}{dt} = -X \tag{1.1}$$

Price-sensitive demand:

$$X = \frac{a}{P^\beta} \tag{1.2}$$

Optimal economic decision making:

$$\frac{dP}{dt} = rP \tag{1.3}$$

The decision equation is reached by considering a tradeoff between a unit of resource produced and sold today versus waiting and doing the same later. If P grows faster than r, the interest rate available for investment of money, then conserving the resource for later sale is profitable and producers will do so – the value of the resource grows faster than money. If, on the other hand, P grows slower than r, then conservation is a bad investment and selling now is preferable – money grows faster than the value of the resource, and a resource owner would prefer to produce now and invest the proceeds at rate r. The price equation expresses the point of indifference between these two options; it would be realized in a situation of competition among many producers. (This is Hotelling's Rule [42]. There will be more to say about this later.)

The solution for P is

$$P = P_0 e^{rt} \tag{1.4}$$

and thus we have the production rate X, from the demand function

$$X = \frac{a}{P_0^\beta} e^{-\beta rt} \tag{1.5}$$

and the initial production rate is

$$X_0 = \frac{a}{P_0^\beta} \tag{1.6}$$

Since $dS/dt = -X$, we have

$$S(t) = S_0 - \int_0^t X dt = S_0 - \frac{a}{P_0^\beta \beta r}\left[1 - e^{-\beta rt}\right] \tag{1.7}$$

We require two conditions to close the system: S_0, the present amount of the resource, and P_0, the initial price.

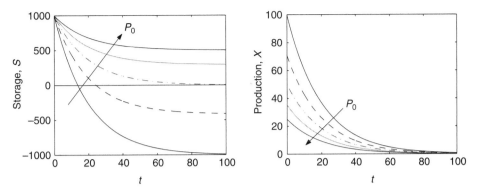

Figure 1.1. Five different depletion histories, identical except for initial price. P_0 increases by factors of 2 in the direction of the arrows.

S_0 is presumed known; P_0 is not. If P_0 is set too high, the demand will be stunted and the resource will go unutilized. If P_0 is set too low, the demand will be too large and the resource will be depleted prematurely, leaving our mathematics of decision making invalid (Figure 1.1). The system is closed by invoking the Terminal Condition (TC): complete resource exhaustion as time goes to infinity:

$$S \to 0 \quad \text{as } t \to \infty \tag{1.8}$$

Thus,

$$S_0 = \frac{a}{P_0^{\beta} \beta r} \tag{1.9}$$

The initial price is thus

$$P_0 = \left[\frac{a}{S_0 \beta r} \right]^{\frac{1}{\beta}} \tag{1.10}$$

and the initial production rate is

$$X_0 = \frac{a}{P_0^{\beta}} = \beta r S_0 \tag{1.11}$$

If P_0 is too high (X_0 too low), then S is never exhausted. If P_0 is too low (X_0 too high), then the resource is exhausted prematurely. In either case, production would be adjusted to satisfy the TC (Figure 1.2).

Rent is the integrated present worth of all net revenues:

$$R = \int_0^{\infty} e^{-rt} X(t) P(t) dt \tag{1.12}$$

Since $P = P_0 e^{rt}$, we have

$$R = P_0 \int_0^{\infty} X(t) dt = P_0 S_0 \tag{1.13}$$

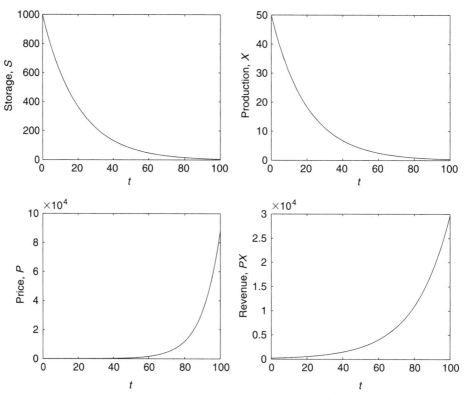

Figure 1.2. Exhaustion history that matches the TC. Demand is $X = a/P^\beta$, with $(a, \beta) = (100, 0.5)$.

For this simple case, the present worth of all future production is equal to today's price times today's total supply.

Program **Oil1** illustrates the exhaustion history under these conditions. The 2 ODE's are integrated forward in time with an explicit (Euler) forward-difference method. The initial price P_0 needs to be adjusted manually to satisfy the TC. Because numerical integration is not perfect, the relations developed above using the calculus correspond only approximately to the **Oil1** simulation; the discrepancies vanish as the numerical timestep Δt becomes infinitesimally small.

1.1.2 Finite Demand

Next, add a ceiling price $\overline{P}$, which limits demand (Figure 1.3). Above this price, customers purchase a substitute product. The previous solution, in which P rises without bound, is invalid. Equations 1.1–1.3 still govern, but the TC needs to be altered. The correct TC in this case is

$$S \to 0 \quad \text{as } P \to \overline{P} \tag{1.14}$$

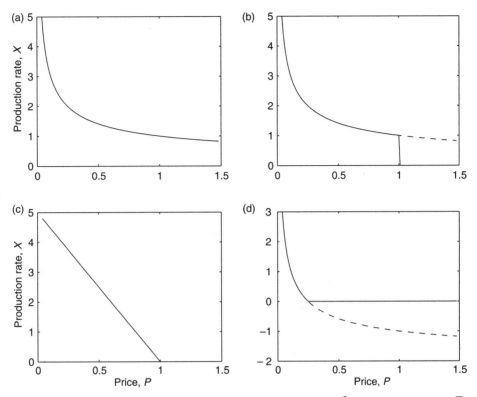

Figure 1.3. Four different demand functions $X(P)$. (a) base case $X_a = P^{-\beta}$; (b) base case with $P \le \overline{P} = 1$; (c) linear demand $X_c = X_0(1 - P)$; (d) base case shifted, $X_d + 2 = P^{-\beta}$. The dash lines indicate the continuation of the base case curve beyond $X \le 0$. Cases b, c, and d have finite demand.

which leads to exhaustion at finite time T. From Equations 1.1 and 1.3, we have

$$\overline{P} = P_0 e^{rT} \tag{1.15}$$

$$S_0 = \frac{a}{P_0^\beta \beta r}\left[1 - e^{-\beta rT}\right] \tag{1.16}$$

from which we obtain the final results

$$P_0 = \left[\frac{a}{\beta r S_0 + \frac{a}{\overline{P}^\beta}}\right]^{\frac{1}{\beta}} \tag{1.17}$$

$$X_0 = \beta r S_0 + \frac{a}{\overline{P}^\beta} \tag{1.18}$$

$$T = \frac{1}{\beta r}\ln\left[\frac{\beta r S_0 \overline{P}^\beta}{a} + 1\right] \tag{1.19}$$

These relations reduce to the previous ones as $\overline{P} \to \infty$.

The above relations must govern at any time during the extraction history, else the trajectory would not be optimal and it would be altered, contrary to hypothesis.

Hence, we may drop the subscripts "0" and it is always true that

$$P = \left[\frac{a}{\beta r S + \frac{a}{\overline{P}^{\beta}}} \right]^{\frac{1}{\beta}} \tag{1.20}$$

$$X = \beta r S + \frac{a}{\overline{P}^{\beta}} \tag{1.21}$$

$$T = \frac{1}{\beta r} \ln \left[\frac{\beta r S \overline{P}^{\beta}}{a} + 1 \right] \tag{1.22}$$

with S the remaining unexploited resource at any time t and P, X, T the contemporary price, production rate, and remaining time to exhaustion. Equations 1.20, 1.21, and 1.22 completely characterize the solution to Equations 1.1, 1.2, and 1.3, subject to the TC of exhaustion as price reaches the ceiling $\overline{P}$.

Program **Oil1a** illustrates these relationships. Figure 1.4 displays simulation results for finite $\overline{P}$ and T, achieved with the decision rule $X = X(S)$ (Equation 1.21).

Rent is, as above, the integrated present worth of all future production:

$$R = \int_0^{\infty} P X e^{-rt} dt = P_0 \int_0^T X dt = P_0 S_0 \tag{1.23}$$

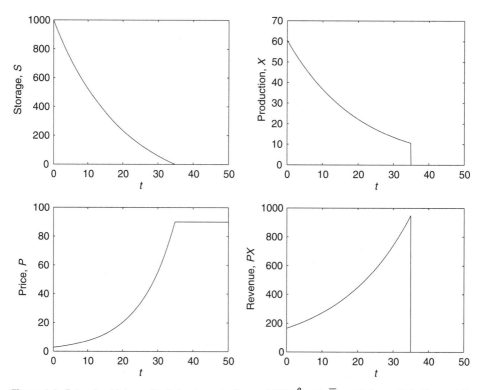

Figure 1.4. Extraction history with finite demand, $X_a = 100 P^{-\beta}$ with $\overline{P} = 90$ (case (b) in Figure 1.3). This leads to exhaustion at finite time as shown.

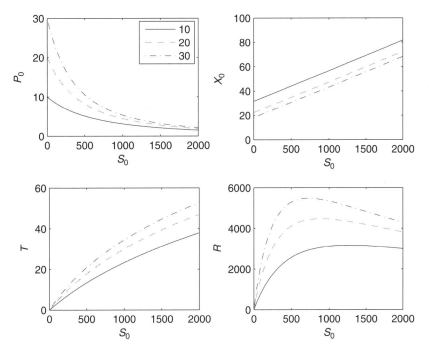

Figure 1.5. P, X, and T as funtions of the total reserve S at any time (Equations 1.20–1.22). Demand: $X = aP^{-\beta}$; $a = 100$; $\beta = 0.5$; $r = 0.05$; $\overline{P} = (10, 20, 30)$ as indicated by the linestyles.

This result is unchanged by the imposition of a ceiling price and the resultant finite T. Since P_0 decreases as $\overline{P}$ decreases, ceiling price has the effect of diminishing overall rent, in accord with intuition.

Equations 1.20–1.22 give P, X, and T as funtions of the total reserve S at any time, assuming complete exhaustion, $X = aP^{-\beta}$ and $P \leq \overline{P}$. Figure 1.5 plots these for three different values of $\overline{P}$. Rent peaks and begins to decline with S at high abundance in this scenario.

Consumers' Surplus

Consumption at price P indicates a willingness to pay at least P – that is, the value of the consumption $V \geq P$. The consumer obtains a surplus equal to the difference $V - P$. Figure 1.6 illustrates a demand curve made up of individual uses ΔX, each with its own value V. If price is set at P, those users with higher value will purchase, and those with lower value will not. The consumers' surplus (CS) is their accumulation:

$$CS = \sum (V - P)\Delta X \tag{1.24}$$

for all increments where $(V - P) > 0$. In the limit,

$$CS(P) = \int_0^{X(P)} (V - P)dx \tag{1.25}$$

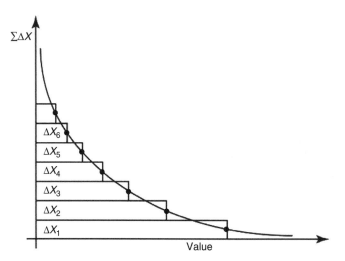

Figure 1.6. Demand curve built up of individual increments ΔX, ordered by decreasing individual value, plotted on the horizontal axis.

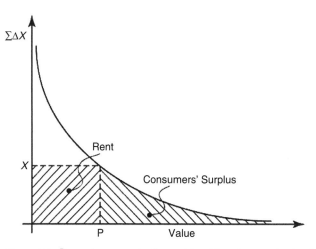

Figure 1.7. Demand curve as in Figure 1.6, adding the actual price P. The area to the right of the price line is the consumers' surplus; that to the left is the rent transferred to the seller.

Clearly, CS is a function of P. Graphically, this is illustrated in Figure 1.7 as the area "under the demand curve, to the right of P". The amount PX shown graphically is the total rent transferred to the seller. So transactions at P generate consumers' surplus as well as rent.

Graphically, it is easy to see that an equivalent integral is

$$\text{CS}(P) = \int_{P}^{\overline{P}} X dV \tag{1.26}$$

An analogous concept of producers' surplus (PS) divides the rent into production cost plus surplus: net rent. When production is costly, the producers' surplus is the net rent.

Consumers' surplus is a static concept; time is fixed in its construction. Clearly in a depletion context, as P rises over time, CS will decrease: $CS(P) = CS(P(t))$. Suppose we have the base case demand $X = aP^{-\beta}$, with a ceiling price $\overline{P}$. Then it is easy to verify that CS may be integrated to obtain

$$CS(P) = \frac{a}{1 - \beta}\left[\overline{P}^{1-\beta} - P^{1-\beta}\right] \tag{1.27}$$

The present worth of the consumers' surplus is explored in Problem 34.

1.1.3 Linear Demand

As an extension of the preceding, consider the alternative demand function

$$P = \overline{P} - bX \tag{1.28}$$

We still have the requirement of exponential price growth

$$P = P_0 e^{rt} \tag{1.29}$$

and thus

$$X = \frac{\overline{P} - P_0 e^{rt}}{b} \tag{1.30}$$

Integrating $dS/dt = -X$ gives

$$S(t) = S_0 + \frac{1}{b}\left[\frac{P_0}{r}(e^{rt} - 1) - \overline{P}t\right] \tag{1.31}$$

The Terminal Condition is

$$S(T) \to 0 \quad \text{as } P(T) \to \overline{P} \tag{1.32}$$

and therefore

$$P_0 e^{rT} = \overline{P} \tag{1.33}$$

$$rT = \ln\left(\frac{\overline{P}}{P_0}\right) \tag{1.34}$$

and

$$S_0 = -\frac{1}{b}\left[\frac{P_0}{r}(e^{rT} - 1) - \overline{P}T\right] \tag{1.35}$$

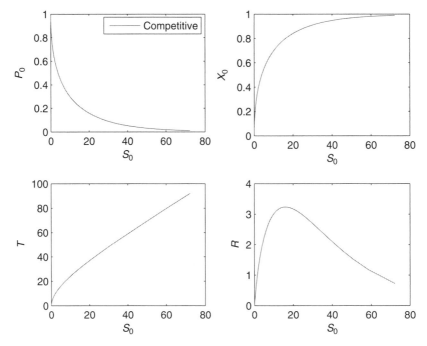

Figure 1.8. Optimal extraction relations for the linear demand case: $r = 0.05$; $P = 1 - X$; competitive case.

A little rearrangement leads to

$$T = \frac{1}{r}\ln\left(\frac{\overline{P}}{P_0}\right) \tag{1.36}$$

$$S_0 = \frac{\overline{P}}{br}\left[\frac{P_0}{\overline{P}} - 1 + \ln\left(\frac{\overline{P}}{P_0}\right)\right] \tag{1.37}$$

$$X_0 = \frac{\overline{P} - P_0}{b} \tag{1.38}$$

These last three equations relate S, X, and T to P_0; they comprise implicit functions $X(S)$, $T(S)$, and $P_0(S)$. There are no simple closed-form solutions, but $X(S)$, $T(S)$, and $P_0(S)$ can be evaluated numerically as in **Oil6M+C**; the plots shown therein are reproduced in Figure 1.8. They characterize this system under linear demand, as did the closed-form Equations 1.20–1.22 for the earlier demand function.

As before, rent $R = P_0 S_0$. It is interesting to note here that as S increases, P_0 ultimately decreases toward the limiting case $P_0 \to 0$ as $S \to \infty$. As a result, the R initially grows with S but ultimately peaks and then decreases with increasing S. The point of maximum rent is found by setting $dR/dt = 0$; the result is

$$\ln\left(\frac{\overline{P}}{P_0}\right) = 2\left(1 - \frac{P_0}{\overline{P}}\right) \tag{1.39}$$

There is one root of this equation in the range $\frac{P_0}{\overline{P}} < 1$: $\frac{P_0}{\overline{P}} \sim 0.2$. The peak value of S is

$$S^* = \frac{\overline{P}}{br}\left[1 - \frac{P_0}{\overline{P}}\right] \sim \frac{\overline{P}}{br}\,[.8] \tag{1.40}$$

More resource beyond this limit results in less overall rent. Consumers' surplus for this case is explored in Problem 35.

The program **Oil6M+C** provides a simulation of X, P, S, and R versus time.

1.1.4 **Expanding Demand**

Going back to the base case of an unbounded demand curve, Equation 1.2, consider the case $a = a(t)$, an exogenous trend toward higher demand. This represents the same general price sensitivity, but growth in a represents an outward shift in demand:

$$X = \frac{a(t)}{P^\beta} \tag{1.41}$$

This is illustrated in Figure 1.9. For example, a simple doubling of the number of resource-consuming machines would be expected to double the demand for the resource, all other things (including P) being constant. Increasing the intrinsic use-per-machine would have the same effect. Growth in a would reflect the product of these two effects, extrinsic and intrinsic resource use or, equivalently, the resource intensity of individual use and the number of uses.

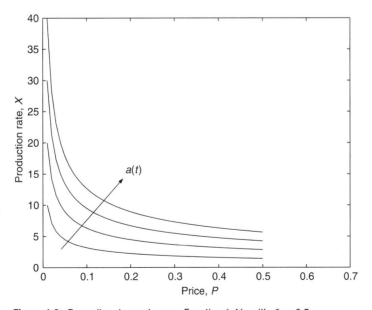

Figure 1.9. Expanding demand curve, Equation 1.41, with $\beta = 0.5$.

As before we have

$$\frac{dS}{dt} = -X \qquad (1.42)$$

$$\frac{dP}{dt} = rP \qquad (1.43)$$

Integrating the P equation gives the price and production history:

$$P(t) = P_0 e^{rt} \qquad (1.44)$$

$$X(t) = \frac{a(t)}{P_0^{\beta}} e^{-\beta rt} \qquad (1.45)$$

Exponential Demand Growth

For this case,

$$a(t) = a_0 e^{gt} \qquad (1.46)$$

$$X(t) = \frac{a_0}{P_0^{\beta}} e^{(g-\beta r)t} = X_0 e^{(g-\beta r)t} \qquad (1.47)$$

With $g < \beta r$, we have an unbounded future of finite production and the TC of resource exhaustion at $T \to \infty$. Integrating gives us

$$S_0 = \frac{a_0}{P_0^{\beta}} \frac{1}{\beta r - g} = \frac{X_0}{\beta r - g} \qquad (1.48)$$

and finally,

$$X_0 = (\beta r - g)S_0 \qquad (1.49)$$

$$P_0 = \left[\frac{a_0}{(\beta r - g)S_0} \right]^{\frac{1}{\beta}} \qquad (1.50)$$

By comparison with the no-growth case, X is initially smaller and decays at the slower rate $e^{(g-\beta r)t}$. Price is initially higher, and its growth rate is unchanged, e^{rt}. Rent $= P_0 S_0$ is higher, proportional to P_0. As the demand growth g approaches βr, pumping becomes infinitesimally small and nearly constant over time; price responds inversely and grows without bound. For $g > \beta r$, the resource is never produced; the owner conserves in the face of rapidly escalating demand.

The case of declining demand (contraction) is the reverse: $g < 0$. In this case, X is initially larger than the constant-demand ($g = 0$) reference case, and its decay rate is faster.

Linear Demand Growth

Next we examine the linear growth case

$$a(t) = a_o + a_1 t \qquad (1.51)$$

with exponential price growth

$$P = P_0 e^{rt} \tag{1.52}$$

Production is again according to the demand curve

$$X = \left[\frac{a_o + a_1 t}{P_0^\beta} \right] e^{-\beta rt} \tag{1.53}$$

At long time, $X \to 0$; but early growth in X is possible. Differentiating,

$$\frac{dX}{dt} = \frac{1}{P_0^\beta} \left[a_1 - \beta r (a_0 + a_1 t) \right] e^{-\beta rt} \tag{1.54}$$

This is positive at $t = 0$ when

$$a_1 > a_0 \beta r \tag{1.55}$$

and if that is the case, there is a maximum in X at t^*:

$$t^* = \frac{(a_1 - a_0 \beta r)}{a_1 \beta r} \tag{1.56}$$

For $t > t^*$, production declines toward zero at $T \to \infty$. For exhaustion at that point, we require

$$S_0 = \frac{a_0}{P_0^\beta \beta r} \left[1 + \frac{a_1}{a_0 \beta r} \right] = \frac{X_0}{\beta r} \left[1 + \frac{a_1}{a_0 \beta r} \right] \tag{1.57}$$

Equivalently,

$$X_0 = \frac{\beta r S_0}{\left[1 + \frac{a_1}{a_0 \beta r} \right]} \tag{1.58}$$

Compared with the no-growth case ($a_1 = 0$), growth introduces a smaller initial production that peaks early, then decays to zero, asymptotically as $te^{-\beta rt}$. (Small demand growth may not produce the peak.) Initial price is higher, growing at the rate e^{rt}.

$$P_0 = \left[\frac{a_0 \left[1 + \frac{a_1}{a_0 \beta r} \right]}{\beta r S_0} \right]^{\frac{1}{\beta}} \tag{1.59}$$

Rent is accordingly higher, $P_0 S_0$.

Figure 1.10 illustrates the production history in this case, for peaking parameters $a_1 > a_0 \beta r$.

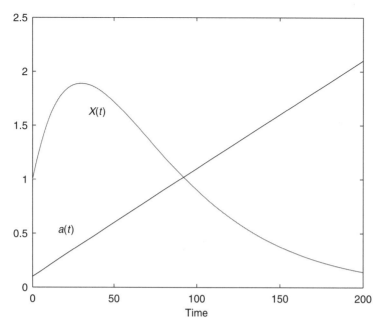

Figure 1.10. Time history of production with linear growth in demand. Parameters: $a_0 = 0.1$; $a_1 = 0.01$; $X_0 = 1.$; $\beta = 0.5$; $r = .05$. Production peaks at $t^* = 30$.

Saturating Demand Growth

Consider the demand function in Equation 1.41 with

$$a(t) = a_0 \left(1 - e^{-st}\right) \tag{1.60}$$

Demand is initially zero and rises toward the saturation level a_0 at the exponential rate s. Again with $P = P_0 e^{rt}$, we have the production rate

$$X = \frac{a_0}{P_0^\beta} \left(1 - e^{-st}\right) e^{-\beta rt} \tag{1.61}$$

and the reserve history

$$S(t) = S_0 - \frac{a_0}{P_0^\beta} \left[\frac{1 - e^{-\beta rt}}{\beta r} - \frac{1 - e^{-(s+\beta r)t}}{s + \beta r} \right] \tag{1.62}$$

Applying the TC of complete exhausition as $t \to \infty$ gives

$$S_0 = \frac{a_0}{\beta r P_0^\beta} \left[\frac{s}{s + \beta r} \right] \tag{1.63}$$

and Equations 1.61 and 1.62 for production and remaining stock become

$$X(t) = S_0 \left[\frac{s + \beta r}{s} \right] \left[1 - e^{-st}\right] e^{-\beta rt} \tag{1.64}$$

$$S(t) = \beta r S_0 \left[\frac{s + \beta r}{s} \right] \left[1 - \left(\frac{\beta r}{s + \beta r} \right) e^{-st} \right] e^{-\beta rt} \qquad (1.65)$$

Starting from zero, demand growth produces early growth in production X, a peak at t^*, and a decline at the rate βr once demand has reached saturation at a_0. The peak production occurs at

$$t^* = \frac{1}{s} \ln \left[\frac{s + \beta r}{\beta r} \right] \qquad (1.66)$$

In the limit of extremely fast saturation (very large $s \to \infty$), we recover the base case: instantaneous onset of production, followed by monotonic decay ($t^* = 0$) at the rate $-\beta r$. For intermediate saturation rates ($s > \beta r$), production at long time approaches the limiting relation $X(t) = \beta r S(t)$, both decaying at the asymptotic rate βr. Essentially, at saturated demand we recover asymptotically the initial base case of constant demand.

When demand development is very slow, $s < \beta r$, then its development dominates production; but the peak production time t^* remains early and insensitive to s. Expanding t^* in a Taylor series gives

$$t^* \to \frac{1}{\beta r} \qquad (1.67)$$

This limit recovers the comparable result for linear demand growth, Equation 1.56, as it should.

Figure 1.11 illustrates the production history for demand saturating exponentially at a finite rate. Figure 1.12 illustrates the effect of s with all other parameters constant, including S_0.

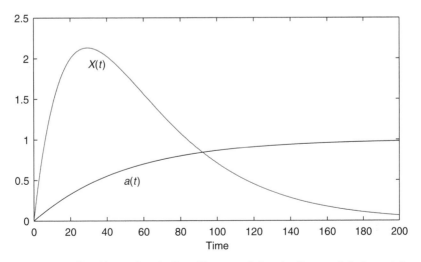

Figure 1.11. Time history of production with exponentially saturating growth in demand. Parameters: $a_0 = 1$; $s = 0.02$; $\beta = 0.5$; $r = 0.05$. Peak production at $t^* = 29.4$.

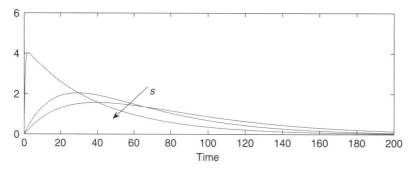

Figure 1.12. Time history of production X for slow, intermediate, and fast growth saturation. Parameters: $a_0 = 1$; $s = 0.0002$, 0.02, and 2.0; $\beta = 0.5$; $r = 0.05$; initial supply fixed at $S_0 = 170$.

Note that this analysis requires knowledge from the beginning that demand will ultimately reach a_0. That knowledge is embedded in the decision making.

Endogenous Demand Growth

It is interesting to speculate on what drives demand growth. We imagine a resource, like oil, which requires capitalization in combustion (fuels) and/or chemical manufacturing (plastics) in order to be used; and these secondary products have their own consumer markets. In these cases, capital formation has an intrinsic lifetime of order 5–20 years (decay rate $\rho = 0.05 - 0.20$). So we speculate that the growth in a would be related to capital formation in the secondary industry:

$$\frac{da}{dt} + \rho a = f(S, X, S/X, aS/X, P, \pi) \qquad (1.68)$$

Increases in a could be driven by perception of large reserves S, large production X, large increases in reserves dS/dt, the exhaustion timescale S/X, changes in P as a metric of scarcity, and/or the rate of rent formation π. Clearly, the dynamics of the secondary industry, and their linkage to the primary resource industry, become critical here. This is beyond our scope.

1.2   DECISION RULES

The decision rule used so far has been

$$\frac{dP}{dt} = rP \qquad (1.69)$$

with r the interest rate applicable to investment of the proceeds of resource sale. This has been applicable to costless production in a competitive market. Three complications need to be explored: taxation, the finite cost of production, and the possibility of monopoly production.

1.2.1 Taxation

If we introduce a tax θ, we have a wedge between the price seen by the consumer P and the net rent $P - \theta$ accruing to the producer. Adapting the principle of intertemporal maximization of net rent (as used above), we have the decision rule

$$\frac{d(P - \theta)}{dt} = r(P - \theta) \tag{1.70}$$

and

$$(P - \theta) = (P_0 - \theta)e^{rt} \tag{1.71}$$

Two cases are interesting. First, consider a *constant tax* θ, which might be conceived as a *consumption tax*. In this case, we have

$$P(t) = \theta + (P_0 - \theta)e^{rt} \tag{1.72}$$

With θ fixed, it loses relevance as P grows. Consumption will be according to the demand function – for example,

$$X(t) = \frac{a}{P^{\beta}} = a\left[\theta + (P_0 - \theta)e^{rt}\right]^{-\beta} \tag{1.73}$$

All other things remaining the same, initial price P_0 will rise with θ, but it will not completely offset it; $P_0 - \theta$ will be less, so that initial net rent is less because of θ. This effect will be less pronounced as production proceeds; ultimately, growth in P overwhelms constant θ.

For example, for $[S_0, r, \beta, a, \Delta t] = [1000, 0.1, 0.5, 100, 0.5]$ and exhaustion at $t \to \infty$: for $\theta = 0$, we find $[P_0, X_0] = [4.4, 47.67]$. The tax case $\theta = 2$ gives $[P_0, X_0] = [5.75, 41.70]$. The early production is delayed by the imposition of the tax; later production is affected less. Program **Oil1-tax** simulates this.

A second case is a *proportional tax*, $\theta = \tau P$. This amounts to a tax on *rent*, not on consumption per se. Following (1.70), we have

$$\frac{dP(1 - \tau)}{dt} = rP(1 - \tau) \tag{1.74}$$

With τ constant, the factor $(1 - \tau)$ cancels, leaving $P = P_0 e^{rt}$ and the whole production history unaffected by τ. The only effect is to redirect a portion of the rent, τR, to the public treasury, leaving the balance $(1 - \tau)R$ for the original resource owner.

1.2.2 Costly Production

Again we generalize the concept of rent here as the present worth of resource sales minus expenses:

$$PX - CX \tag{1.75}$$

where P and X are the price and extraction rate as before and C is the unit cost of production. We will consider the case where C depends on the remaining stock of resource, $C = C(S)$. This would reflect the case where production becomes more costly as extraction proceeds, with C increasing as S decreases. The contributions to rent R from two adjacent production periods separated in time by Δt are

$$R = \left\{ [P_1 - C(S_1)]\, X_1 + \frac{1}{1 + r\Delta t}\, [P_2 - C(S_2)]\, X_2 \right\} \tag{1.76}$$

We wish to discover the effect of adjusting a hypothetical production pattern by moving a quantity Δ of production forward in time, at the expense of later production, leaving all other things the same. We introduce the perturbations

$$\begin{aligned}
\Delta X_1 &= \Delta \\
\Delta X_2 &= -\Delta \\
\Delta S_1 &= 0 \\
\Delta S_2 &= -\Delta \cdot \Delta t
\end{aligned} \tag{1.77}$$

Perturbing R with the changes ΔX and ΔS, we obtain

$$\Delta R = \left\{ [P_1 - C(S_1)]\, \Delta X_1 - X_1 \frac{dC}{dS} \Delta S_1 + \frac{1}{1 + r\Delta t} \left[[P_2 - C(S_2)]\, \Delta X_2 - X_2 \frac{dC}{dS} \Delta S_2 \right] \right\} \tag{1.78}$$

With the perturbations above, we obtain

$$\Delta R = \left\{ [P_1 - C(S_1)]\, \Delta - \frac{1}{1 + r\Delta t} \left[[P_2 - C(S_2)]\, \Delta - X_2 \frac{dC}{dS} \Delta \cdot \Delta t \right] \right\} \tag{1.79}$$

The point of indifference to such an adjustment is $\Delta R = 0$:

$$[P_1 - C(S_1)] = \frac{1}{1 + r\Delta t} \left[[P_2 - C(S_2)] - X_2 \frac{dC}{dS} \Delta t \right] \tag{1.80}$$

Rearranging this, we obtain

$$[P_2 - C(S_2)] = [1 + r\Delta t]\, [P_1 - C(S_1)] + X_2 \frac{dC}{dS} \Delta t \tag{1.81}$$

and taking the limit as $\Delta t \to 0$, we have

$$\frac{d(P - C)}{dt} = r(P - C) + X \frac{dC}{dS} \tag{1.82}$$

Here we find two effects:

1. The rent rate $(P - C)$, not price alone, must generally increase at the rate r with time.

2. There is an additional effect of variable costs, $X dC/dS$, called the "stock effect." With $X > 0$ and $dC/dS < 0$ (C increasing as resource exhaustion proceeds), this term will be negative. Thus, the rate of rise in $(P - C)$ will be slowed by the stock effect.

Finally, we can rearrange this equation to a more convenient form by noting that $-X = dS/dt$ and therefore by the chain rule,

$$X \frac{dC}{dS} = -\frac{dC}{dS} \frac{dS}{dt} = -\frac{dC}{dt} \tag{1.83}$$

Thus, we have the simpler form of the **decision rule under costly production:**

$$\frac{dP}{dt} = r(P - C) \tag{1.84}$$

Again it is clear that as C becomes significant, it slows the exponential growth that would characterize P in the costless production case. As the resource dwindles, we expect growth in C to bring rent toward zero, with growth in P slowing and P finally stabilizing at $\overline{P}$. The TC becomes $R \to 0$, $P \to \overline{P}$, $C \to \overline{P}$, and $S \to \underline{S}$, where $\underline{S}$ is the small residual left in storage that is not ecomonical to extract:

$$C(\underline{S}) = \overline{P} \tag{1.85}$$

Naturally, these relations reduce to the costless case studied earlier when $C = 0$.

Program **Oil2** simulates the simple case of constant cost – for example, a tax on production. In this case, we have the simple rule

$$\frac{d(P - C)}{dt} = r(P - C) \tag{1.86}$$

that is, there is no stock effect. Demand remains sensitive only to P alone, so the solution is not simply to shift P from the costless case. **Oil4** simulates the stock effect with cost function

$$C(S) = \gamma / S^{\delta} \tag{1.87}$$

1.2.3 Monopoly versus Competitive Production

Next we turn to the monopoly case, wherein a single producer supplies the whole market. Under this condition, production decisions directly affect price; while under competetive conditions, the effect is only via the aggregate of several producers' independent decisions. The effect is on the revenue from sales PX. Perturbing this product gives two terms:

$$\Delta(PX) = P\Delta X + X\Delta P \tag{1.88}$$

And with the demand function $P = P(X)$, we have

$$\Delta(PX) = P\Delta X + X\frac{dP}{dX}\Delta X \tag{1.89}$$

Defining the demand **elasticity** $\epsilon \equiv \frac{X}{P}\frac{dP}{dX}$, we have

$$\Delta(PX) = P(1 + \epsilon)\Delta X \tag{1.90}$$

Redoing the perturbation analysis in the previous section, we obtain the additional elasticity terms as follows:

$$\frac{dP(1 + \epsilon)}{dt} = r\left[P(1 + \epsilon) - C\right] \tag{1.91}$$

which is the **decision rule under costly monopoly production**. (The reader should check this.)

For the demand function

$$X = \frac{a}{P^{\beta}} \tag{1.92}$$

we have

$$\epsilon = -\frac{1}{\beta} \tag{1.93}$$

that is, a constant; for this demand function, monopoly and competitive producers would behave the same. For the linear demand function

$$P = \overline{P} - bX \tag{1.94}$$

we have

$$\epsilon = -\left(\frac{\overline{P}}{P} - 1\right) \tag{1.95}$$

Thus, in this case ϵ is negative and increases toward 0 as P increases toward $\overline{P}$.

In the monopoly case, it is convenient to introduce the variable $Q \equiv P(1 + \epsilon)$. The equation

$$\frac{dQ}{dt} = r(Q - C) \tag{1.96}$$

then replaces its competitive version with P replacing Q. The example below illustrates this.

Monopoly Production under Linear Demand (Costless)

As an example, consider the linear demand case

$$P = \overline{P} - bX \tag{1.97}$$

for which we have

$$\epsilon = -\left(\frac{\bar{P}}{P} - 1\right) \qquad (1.98)$$

This case can be integrated in closed form as follows. First, introduce the quantity $Q \equiv P(1 + \epsilon)$. In the present case,

$$Q = 2P - \bar{P} \qquad (1.99)$$

and equivalently

$$P = \frac{1}{2}(Q + \bar{P}) \qquad (1.100)$$

Since $dQ/dt = rQ$, we have

$$Q = Q_0 e^{rt} \qquad (1.101)$$

and from the demand function,

$$X = \frac{\bar{P} - Q_0 e^{rt}}{2b} \qquad (1.102)$$

Integrating $\frac{dS}{dt} = -X$ gives

$$S(t) = S_0 - \frac{1}{2b}\left[\bar{P}t - \frac{Q_0}{r}\left(e^{rt} - 1\right)\right] \qquad (1.103)$$

The various Initial Conditions are

$$\begin{aligned} Q_0 &= 2P_0 - \bar{P} \\ P_0 &= \frac{1}{2}(Q_0 + \bar{P}) \\ X_0 &= \frac{1}{b}(\bar{P} - P_0) \end{aligned} \qquad (1.104)$$

The Terminal Condition requires $S(T) \to 0$ as $P(T) \to \bar{P}$, and therefore we have

$$\begin{aligned} S(T) &= 0 \\ P(T) &= \bar{P} \\ Q(T) &= \bar{P} \end{aligned} \qquad (1.105)$$

and since $Q(T) = Q_0 e^{rT}$,

$$rT = \ln\left(\frac{Q(T)}{Q_0}\right) \qquad (1.106)$$

After some manipulations, we arrive at the final set of relations among (S_0, X_0, T) and P_0:

$$T = \frac{1}{r} \ln \left(\frac{\overline{P}}{2P_0 - \overline{P}} \right) \tag{1.107}$$

$$S_0 = \frac{1}{rb} \left[\frac{\overline{P}}{2} \ln \left(\frac{\overline{P}}{2P_0 - \overline{P}} \right) + (P_0 - \overline{P}) \right] \tag{1.108}$$

$$X_0 = \frac{1}{b} (\overline{P} - P_0) \tag{1.109}$$

This solution requires $P_0 > \overline{P}/2$.

The reader is encouraged to compare these relations with the comparable results for competitive production in Equations 1.36–1.38 (Problems 6 and 7). Generally, monopoly production results in higher initial prices, longer exhaustion times, and higher overall rent than competitive production. There would be an incentive for competitive producers to form a monopoly and share the extra rent generated. It is interesting that this leads to slower overall resource exhaustion.

Total rent under this scenario is

$$
\begin{aligned}
R &= \int_0^\infty P(t)X(t)e^{-rt}\,dt \\
&= \int_0^T \frac{(Q_0 e^{rt} + \overline{P})}{2} \frac{(\overline{P} - Q_0 e^{rt})}{2b} e^{-rt}\,dt \\
&= \frac{1}{4b} \int_0^T (\overline{P}^2 - Q_0^2 e^{2rt})e^{-rt}\,dt \\
&= \frac{1}{4b} \int_0^T (\overline{P}^2 e^{-rt} - Q_0^2 e^{rt})\,dt \\
&= \frac{1}{4rb} (\overline{P}^2 [1 - e^{-rT}] + Q_0^2 [1 - e^{rT}])
\end{aligned}
\tag{1.110}
$$

With some effort, one can simplify this as follows. Using

$$1 - e^{-rT} = 2 \left(\frac{\overline{P} - P_0}{\overline{P}} \right) \quad \text{and} \quad 1 - e^{rT} = 2 \left(\frac{P_0 - \overline{P}}{2P_0 - \overline{P}} \right) \tag{1.111}$$

the final result becomes

$$R = \frac{1}{br} (\overline{P} - P_0)^2 \tag{1.112}$$

Monopoly rent in this case is clearly always nonnegative. There is no peak at intermediate S, unlike the competitive case illustrated in Figure 1.8.

Figure 1.13 illustrates these relations among X_0, T, P_0, S_0, and R for monopoly production. The competitive case plotted in Figure 1.8 is replicated for comparison. **Oil6M+C** provides a simulation of this system's time evolution.

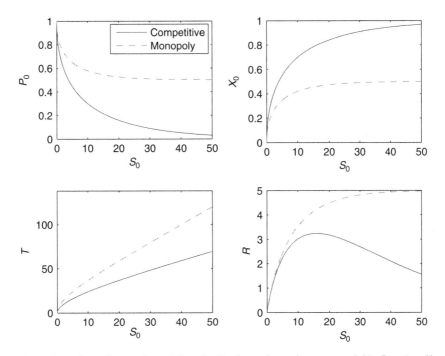

Figure 1.13. Optimal extraction relations for the linear demand case: $r = 0.05$; $P = 1 - X$; monopoly case (dash line). The competitive case (solid line) is replicated from Figure 1.8 for comparison.

1.3 DISCOVERY

Next we add an inventory of undiscovered resource U to the picture and the rate of discovery D:

$$\frac{dS}{dt} = -X + D \tag{1.113}$$

$$\frac{dU}{dt} = -D \tag{1.114}$$

$(S + U)$ is the total unproduced resource; initial values are S_0 and U_0. To close the equations, we need expressions for X and D.

1.3.1 Exogenous Discovery

Consider first the simple case where discovery is exogenous – that is, it is completely independent of P, X, and S. A example would be the first-order discovery rate

$$D = \rho U \tag{1.115}$$

This discovery rate is proportional to the remaining undiscovered resource; it is high when there is a lot to find, and it decreases as the discovery history proceeds and the remaining U decreases.

For production, we invoke the result of the previous analyses: With demand $X = a/P^\beta$, ceiling price $\overline{P}$, and zero production cost, we have

$$X = \beta r S + \frac{a}{\overline{P}^\beta} \tag{1.116}$$

Essentially, we assume the production decision is pessimistic, based on a forecast of zero future discovery. This production relation is rational (optimal) according to the previous analyses.

The complete system is now

$$\frac{dS}{dt} + \beta r S = \rho U - \frac{a}{\overline{P}^\beta} \tag{1.117}$$

$$\frac{dU}{dt} + \rho U = 0 \tag{1.118}$$

The solution for U is straightforward:

$$U = U_0 e^{-\rho t} \tag{1.119}$$

and for S, we now have

$$\frac{dS}{dt} + \beta r S = \rho U_0 e^{-\rho t} - \frac{a}{\overline{P}^\beta} \tag{1.120}$$

The solution for S will have two parts: a "homogeneous" part of the form $e^{-\beta r t}$, as in the cases without discovery; and a "particular" part of the form dictated by the right-hand side, $e^{-\rho t}$ plus a constant:

$$S(t) = A e^{-\beta r t} + B e^{-\rho t} + C \tag{1.121}$$

Plugging 1.121 into 1.120, we obtain

$$A[-\beta r + \beta r] e^{-\beta r t} + B[-\rho + \beta r] e^{-\rho t} + C[\beta r] = \rho U_0 e^{-\rho t} - \frac{a}{\overline{P}^\beta} \tag{1.122}$$

The first term vanishes identically. The balance of the equation requires

$$B = U_0 \left[\frac{\rho}{\beta r - \rho} \right] \tag{1.123}$$

$$C = \left[-\frac{a}{\beta r \overline{P}^\beta} \right] \tag{1.124}$$

and the Initial Condition requires $S_0 = A + B + C$, so

$$A = S_0 - B - C \tag{1.125}$$

The complete solution is

$$S(t) = [S_0 - B - C] e^{-\beta rt} + Be^{-\rho t} + C \tag{1.126}$$

Two cases are interesting. If production is fast relative to discovery, then S decays monotonically, with the slow discovery persisting and extending production over a longer time than in the no-discovery case. The criterion for this case is $dS/dt < 0$ at $t = 0$. This requires

$$\frac{\beta r(S_0 - C)}{\rho U_0} > 1 \tag{1.127}$$

If, on the other hand, initial discovery is fast relative to production, then S rises initially. In this case, there will be a peak in S at intermediate time, after which discovery is largely over, production dominates, and S declines. It is easy to find the peak time t_p by setting dS/dt to zero. The result for the case $C = 0$ ($\overline{P} = \infty$) is

$$t_p = \left(\frac{1}{\rho - \beta r}\right) \ln \left[\frac{\rho}{\beta r \left(1 - \frac{S_0}{B}\right)} \right] \tag{1.128}$$

(The student should verify this and develop the case $C \neq 0$.) Beyond $t = t_p$, we are essentially in a depletion phase with discovery largely over.

For example: with $r = 10\%$ per year, $\rho = 10\%$ per year, $\beta = .5$, $\overline{P} = \infty$, $S_0 = 10$, and $U_0 = 10$, we have S peaking at 11.25, at $t_p = 5.75$ years. If, instead, $U_0 = 50$, then S peaks at 30.26 and t_p occurs at 11.96 years. These are illustrated in Figure 1.14. Figure 1.15 illustrates the slower discovery rate, $\rho = 0.04$, which does not exhibit a peak. The Excel and Matlab programs **Oil5a** simulate this system.

More-complex discovery relations are interesting, and their solution is possible via simulation. For example, the price-sensitive discovery rate

$$D = \rho U \left(\frac{P}{P_0}\right)^\gamma \tag{1.129}$$

simulates enhanced exploration effort as the resource scarcity causes price to increase. This may be examined in **Oil5a**. Additional features include the introduction of stochastic disturbances to the otherwise smooth discovery rate. That is available in **Oil5aRandom**.

1.3.2 Discovery Rate: Effort and Efficiency

It is useful to pick apart the various factors influencing the discovery rate: the discovery efficiency; the discovery effort; and the amount left to be discovered.

The efficiency of discovery embodies accumulated geological knowledge, including both where not to look (already looked there) and what features to look for. Thus, we expect a monotonically rising curve as discovery proceeds – a learning curve,

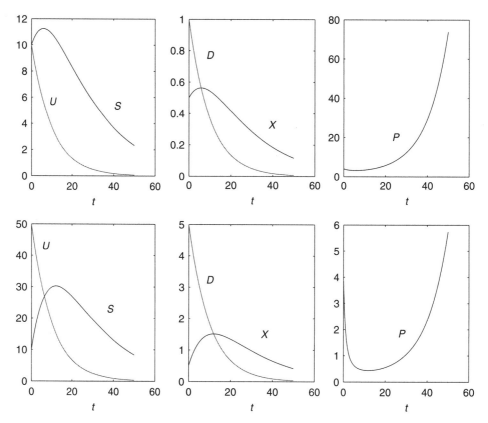

Figure 1.14. Discovery and extraction history, as in the example

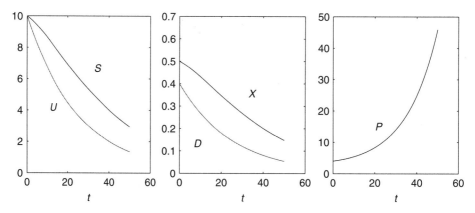

Figure 1.15. Discovery and extraction history, as in Figure 1.14; but ρ is slower, changed from 0.10 to 0.04.

for example, of the form $\left[1 - e^{-(U_0 - U)/U^*}\right]$, with $U_0 - U$ the cumulative discovery and U^* a scaling factor. In addition, we have the inescapable effect of technological invention of devices, sensors, information bases, etc., which we expect will only rise over time in their amplification of discovery rate by the exogenous multiplier $A(t)$.

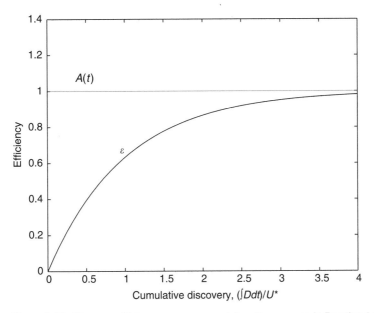

Figure 1.16. Discovery efficiency ϵ versus cumulative discovery, as in Equation 1.130.

The composite of these leads to the hypothetical efficiency function

$$\epsilon = A(t) \left[1 - e^{-(U_0 - U)/U^*} \right] \tag{1.130}$$

This relation is sketched in Figure 1.16. Saturation of the learning curve at its technological limit A occurs when cumulative discovery reaches $4U^*$.

The effect of discovery effort E is monotonically related to the discovery rate. At low levels of effort, the relationship should be linear. At higher levels of effort, we expect decreasing returns to scale due to crowding and duplication of effort. An example of such an effort function, $f(E)$, is

$$f(E) = \left[1 - e^{-E/E^*} \right] \tag{1.131}$$

where E^* is a scaling factor. Like the learning curve above, it saturates at $E \approx 4E^*$. This function is illustrated in Figure 1.17.

These relations together give us the discovery rate D:

$$D = \epsilon \cdot f(E) \cdot U \tag{1.132}$$

At low effort, $E \ll E^*$, we have the linear relationship

$$f(E) \simeq E/E^* \tag{1.133}$$

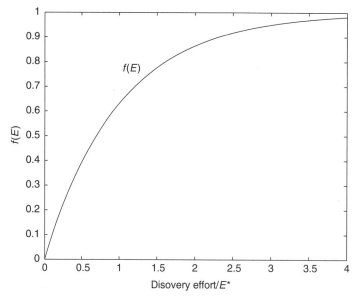

Figure 1.17. Discovery effort multiplier $f(E)$, an increasing function of effort, as in Equation 1.131. This is the effect of effort on discovery rate.

and thus

$$D \simeq \frac{\epsilon}{E^*} EU \qquad (1.134)$$

1.3.3 Effort Level and Exploration Profit

So far, we have a dynamical system involving proven and undiscovered reserves S and U, the production rate X, which is closed in terms of S, and the discovery rate D, which is closed in terms of discovery effort E.

$$\frac{dS}{dt} = -X + D \qquad (1.135)$$

$$\frac{dU}{dt} = -D \qquad (1.136)$$

$$X = \beta r S + \frac{a}{P^\beta} \qquad (1.137)$$

$$D = \epsilon \cdot f(E) \cdot U \qquad (1.138)$$

What sets the level of effort?

To fix ideas, imagine an exploration company selling rights to new discoveries, at the price p_d, to production companies. What is at stake in selling such a new discovery is the present worth of all future rent that could be derived from producing that resource; the rate of rent transfer to the selling company would be $p_d \cdot D$ if a competitive market existed among buyers. (In a costless production economy, p_d would be equal to the market price P of produced resource; otherwise, it would be less.)

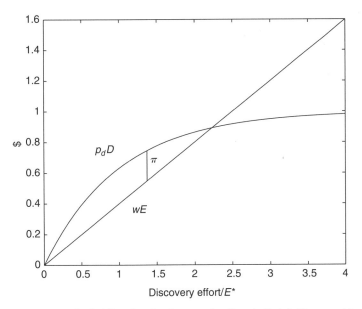

Figure 1.18. Profitability of exploration as a function of effort E. The curved line is the value of discovery; the straight line is its cost. The profitability π is the difference. Parameters: $p_d \epsilon U = 1.$; $wE^* = .4$.

Assume that the cost of exploration is proportional to effort, wE. Then we have the rate of profit earned by exploration as

$$\pi = p_d D - wE = p_d \epsilon f(E) U - wE \qquad (1.139)$$

This is illustrated in Figure 1.18. Competition in the exploration business would drive the effort to its maximum profitable rate, where $\pi = 0$:

$$E = \frac{p_d}{w} D \Rightarrow \frac{f(E)}{E} = \frac{w}{p_d \epsilon U} \qquad (1.140)$$

This is the "open access," or "free entry," situation. All effort is free to mobilize and enter the exploration competitively. It does so until profitability becomes zero.

If, on the other hand, discovery effort could be controlled, then effort would be limited to the point of maximum π, where $d\pi/dE = 0$:

$$\frac{df}{dE} = \frac{w}{p_d \epsilon U} \qquad (1.141)$$

This is the "controlled access," or "restricted entry," case. A monopoly on exploraton would presumably keep effort at this lower level. For the parameters illustrated in Figure 1.18, visual inspection shows that $E \approx 2.2E^*$ for the open, competitive case, and $E \approx .75E^*$ for the controlled-access case. Competition in exploration gives higher effort, faster discovery, and therefore early production.

1.3.4 Effort Dynamics

How does E change with time? In the competitive case, it is reasonable to postulate that rent attracts effort. Denoting by v the first-order rate constant for this process, we have the 3-state system:

$$\frac{dE}{dt} = v\pi \tag{1.142}$$

$$\frac{dS}{dt} = -X + D \tag{1.143}$$

$$\frac{dU}{dt} = -D \tag{1.144}$$

$$X = \beta rS + \frac{a}{P^{\beta}} \tag{1.145}$$

$$D = \epsilon \cdot f(E) \cdot U \tag{1.146}$$

$$\pi = (p_d D - wE) \tag{1.147}$$

$$P = \left(\frac{a}{X}\right)^{\frac{1}{\beta}}; \quad P(t) \leq \overline{P} \tag{1.148}$$

The Excel program **Oil7** simulates these dynamics, under the assumption of cost-less production ($p_d = P$, the market value of production), efficiency saturation

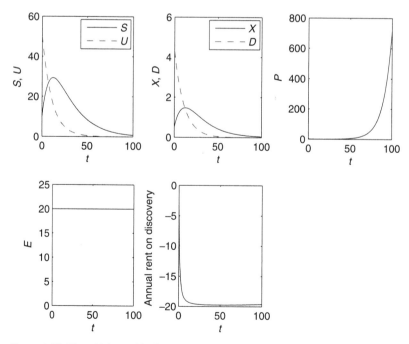

Figure 1.19. Time history with discovery adjusted as in Equations 1.142–1.148, with constant effort. Parameters: $r = .10$; $a = 1$; $\beta = .5$; $\overline{P} = \infty$; $\epsilon = .1$; $E^* = 10$; $p_d = P(t)$; $w = 1$; $v = 0$.

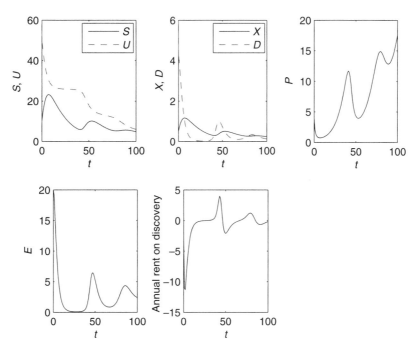

Figure 1.20. Same as in Figure 1.19, except rent-sensitive discovery, $\nu = 0.3$.

(extensive geological knowledge and exploration experience), and constant technology. For small ν, there is an orderly approach to quasi-equilibrium wherein exploration proceeds gradually as S is exhausted. Faster effort response (higher ν) generates a boom-or-bust cycle in exploration, with periodic episodes of discovery followed by layoffs in the exploration business and simple depletion of known reserves as production proceeds. When price rises sufficiently to reinvigorate exploration effort, the cycle is repeated.

An example is presented in Figures 1.19 and 1.20. In Figure 1.19, we plot the case $\nu = 0$, wherein the initial effort $E = 20$ is held constant. Discovery proceeds at a constant proportional rate. The production floods the market, suppressing the price. The exploration is not profitable, but there is no adjustment. Figure 1.20 is the same case, but with $\nu = 0.3$. The initial unprofitable level of effort is reduced quickly, essentially shutting down exploration until roughly $t = 40$, at which point it is profitable to re-initiate exploration effort. There is a complex cycling of effort.

1.4 SOME UNCLOSED ISSUES

The development above is a framework for examining the exhaustible resources. Several questions remain:

- Demand changes reflecting more uses or changes in efficiency; substitution dynamics; technical progress; changes in market participation in the resource consumption. What is da/dt?

- Discovery and the effect of technical progress. What is dA/dt?
- What is the effect of production capacity, the maximum X and dX/dt which is possible? Clearly, that puts a constraint on short-term changes in production.
- The cost of extraction generally: what is $C(S)$?
- How is rent distributed?
- The role of uncertainty, error, bogus data, and uncontrolled stochastic disturbances, generally, on resource management.

Each of these items deserves careful scrutiny in more-elaborate developments.

1.5 RECAP

As mentioned at the outset, this is an extreme case; the example used (petroleum) is not meant to be definitive, only illustrative and helpful to the exposition. Several key ideas emerge, including the role of the owner in setting the decision context and "collecting the rent"; the role of the demand funtion in representing value and "consumers' surplus"; the role of finite demand and substitution; the dynamic of public policy in conditioning decision making as well as rent dispersal; the role of discovery; and the view of exploration as a subsidiary industry. From a sustainability perspective, this resource has only one fate: exhaustion over a finite interval. The sustainable substitute needs to be found during the process: That can be the accumulated rent, treated as a renewable asset; the knowledge gained during the process of exhaustion ("learning how do do without"); the capitalization of a substitute; and/or the social progress achieved during resource exhaustion. Ultimately, substitution alone is inadequate if it necessarily progresses through a series of finite resource exhaustions. Without a sustainable finale, one simply exhausts all resources, irreversibly, leaving no opportunity.

There are several introductory texts on natural resource economics, including Conrad and Clark [13], Neher [68], and Conrad [12]. Griffin [35] covers water resources, which is a necessary supplement for natural resources generally. Conrad [12] includes a valuable annotated bibliography.

All of the above lead to the more general coverage of public goods – for example, Musgrave and Musgrave [66] and Cornes and Sandler [14]. Contemporary scholarship is reemphasizing the foundational notion of merit goods introduced by Musgrave (Ver Eecke [22]; Meier [63]). There are several works that have incorporated globalization as a starting point, notably the work of Kaul et al. [49, 47, 48]. Further study in all of these works is encouraged.

1.6 PROGRAMS

The following programs illustrate the ideas in these lectures. In most cases, there are .xls and .m versions:

- **Oil1:** Costless production, constant-elasticity demand. P_0 is adjustable to meet TC.
- **Oil1a:** Costless production, constant-elasticity demand. Simulation with $X = X(S)$.
- **Oil1-tax:** Costless production, constant-elasticity demand, constant tax.
- **Oil2:** Constant-cost production.
- **Oil4:** Variable-cost production: $C(S) = \gamma/S^\delta$.
- **Oil5a:** Simulates price-sensitive discovery and production.
- **Oil5aRandom:** Adds stochastic perturbation to discovery rate in **Oil5a**.
- **Oil6Monop:** Simulates costless production with linear demand; monopoly production.
- **Oil6Cmptn:** Simulates costless production with linear demand; competitive production.
- **Oil6M+C_Rent:** Combines the two **Oil6** programs; adds the calculation of P_0, X_0, T and the total present worth of rent, as functions of S_0 for the two cases monopoly and competitive production.
- **Oil7:** Like **OiL5a** but adds exploration effort as an endogenous state variable. Discovery rate is the product of fixed efficiency, an increasing function of exploration effort, and remaining undiscovered resource. Effort growth is proportional to the current profitability of exploration.

1.7 PROBLEMS

1 Costless production of a sterile resource:

$$\frac{dS}{dt} = -X \tag{1.149}$$

$$\frac{dP}{dt} = rP \tag{1.150}$$

$$X = \frac{a}{P^\beta} \tag{1.151}$$

where S is the amount in storage, X is the annual production, and P is the price. Simulate the evolution of this system using the following parameters:

$S_0 = 1,000$

$\beta = 0.5$

$r = .10/\text{year}$

$a = 100.$

Plot S, P, and X as functions of time. Confirm that P grows at the rate r and that S decays at the rate βr. By trial and error, confirm that the initial price that exhausts the resource at infinite time is

$$P_0 = \left[\frac{a}{S_0 \beta r} \right]^{\frac{1}{\beta}} \tag{1.152}$$

2 Derive Equations 1.17, 1.18, and 1.19, which summarize the solution with finite demand capped at the substitution price $\overline{P}$.

3 Repeat Problem 1 with $\overline{P} = 12$. What is the Terminal Condition, and what is the theoretical value of P_0 that achieves it? Does that work in the simulation? Compare results with the simulation of Problem 1, wherein $\overline{P}$ is unbounded.

4 The same as Problem 3, but add a sudden, unanticipated reduction of $\overline{P}$ from 12 to 8 at time = 10 years, due to unanticipated technical progress at $t = 10$. Clearly describe what changes need to occur at $t = 10$ and say why.

5 Redo the simulation in Problem 3, adding the unit cost of production C (dollars per unit produced):

$$C = \frac{\gamma}{S^{\frac{1}{3}}} \tag{1.153}$$

such that annual revenues are $= (P-C)X$. Study an interesting range of the parameter γ, ranging from 10 to 100. (Begin with $\gamma = 50$.) Describe how you obtained the initial price and how you altered the equations. Compare with the Problem 3 simulation in terms of initial price, time to exhaustion, production rate, and any other relevant quantities.

6 Costless production of a sterile resource:

$$\frac{dS}{dt} = -X \tag{1.154}$$

$$\frac{dP}{dt} = rP \tag{1.155}$$

$$X = \frac{5}{P} \tag{1.156}$$

where S is the amount in storage, X is the annual production, P is the price, and r is the interest rate.

Today, $S = 400$. There is no substitute for this resource. Interest rate $r = 0.10$ per year.

   (a) What is today's price?
   (b) At $t = 10$, what will the price be? What amount of resource will remain in storage?
   (c) At $t = 10$, there is a new discovery of 300 units of resource. What will the price change to?
   (d) Sketch the time history of P, S, and X, from $t = 0$, until the resource is exhausted.

7 Costless production of a sterile resource; with the linear demand function

$$P = \overline{P} - bX \tag{1.157}$$

In the text, we solve analytically for S, X, and P as functions of time under *competitive* extraction. Confirm that analysis by simulation using $\overline{P} = 1$ and $b = 1$ and P_0, which meets the appropriate Terminal Condition.

We also have expressions for S_0, X_0, and T as functions of the initial price P_0. Evaluate these functions over a useful range of P_0, using $\overline{P} = 1$ and $b = 1$. Use these evaluations to plot the implicit functions X_0, T, and P_0 as functions of the initial storage S_0.

8 Costless competitive production of a sterile resource with finite demand:

$$X + K = \frac{a}{P^\beta} \tag{1.158}$$

This is illustrated in Figure 1.3.
 (a) What is the ceiling price $\overline{P}$?
 (b) Confirm the following relations:

$$S_0(T) = \frac{K}{\beta r}\left[e^{\beta rT} - 1\right] + KT \tag{1.159}$$

$$X_0(T) = K\left[e^{\beta rT} - 1\right] \tag{1.160}$$

$$P_0(T) = \frac{a}{Ke^{\beta rT}} \tag{1.161}$$

At any time, these must relate S, X, and P to the remaining life T of the resource.
 (c) Plot S, X, and P versus T.
 (d) Using the same relations, plot X, P, and T versus S, the amount of resource remaining.

9 Repeat Problem 7 but use *monopoly* production. Compare the results with Problem 7. Do you find that monopoly production is more or less conservative than competitive production?

10 Costless production of a sterile resource is governed by

$$\frac{dS}{dt} = -X \tag{1.162}$$

$$\frac{dP}{dt} = rP \tag{1.163}$$

$$X = \frac{a}{P^\beta} \tag{1.164}$$

where S is the amount in storage, X is the annual production, and P is the price. Parameters are: $a = 20$; $\beta = 0.5$; and $r = 0.1$. Today ($t = 0$), $S = 100$.

(a) Calculate P, S, and X at $t = 0$ and at $t = 5$.

(b) At $t = 5$, new laws are enacted that will regulate and tax production. The new laws will become effective in 10 years, that is, at $t = 15$. Details are unclear; but it is highly likely that production will be either unprofitable, illegal, or both, beginning at $t = 15$. What adjustments to P and X will occur and why? Calculate the new values of P and X following this adjustment.

(c) Sketch the time history of P, X, and S from $t = 0$ to $t = 15$ and beyond.

11 Costless production of a sterile resource, $S_0 = 1{,}000$. Demand is

$$X = \frac{1}{\sqrt{P}} \tag{1.165}$$

subject to the ceiling price $\overline{P} = 15$. Interest rate is $r = .05$.

(a) What are today's price P? production rate X? time to exhaustion T?

(b) A new substitute is about to be announced; the net effect will be that demand will begin decaying over time:

$$X = \frac{e^{-0.1t}}{\sqrt{P}} \tag{1.166}$$

What changes will occur in P, X, and T as soon as this substitute is announced?

(c) What is the value of the research that led to the substitute? Who is happy, who is sad?

12 Costless, competitive production of a sterile resource. Demand is the standard used frequently in class:

$$X = \frac{100}{\sqrt{P}} \tag{1.167}$$

Owner is conservative in the face of discovery. Interest rate is 0.10. Price is observed to be constant over many years, at $P = 100$.

(a) What is the discovery rate?

(b) What is the amount of reserves (S) today?

13 For the case summarized in Equations 1.20–1.23: At what initial abundance S^* does the present worth of rent peak? Obtain this as a function of the parameters $r, a, \beta, \overline{P}$. Confirm your analysis graphically that rent decreases with S for $S > S^*$.

14 Costless, competitive production of a sterile resource. The demand function is given by

$$X = \frac{1}{b} \ln\left(\frac{\overline{P}}{P}\right) \tag{1.168}$$

or equivalently,

$$P = \overline{P}e^{-bX} \tag{1.169}$$

Find the initial price P_0 and the time to exhaustion T, both as functions of the initial storage S_0.

15 Costless production of a sterile resource, with demand given by

$$P = \bar{P} - bX \tag{1.170}$$

Price today is $.6\bar{P}$. The interest rate is 10%. There is no discovery. Predict the price in one year under
  (a) monopoly production
  (b) competitive production

16 Costless, competitive production of a sterile resource. Demand is

$$X = \frac{100}{\sqrt{P}} \tag{1.171}$$

with no price ceiling, interest rate $r = .10$, and total reserves today $S_0 = 1,000$.
It is widely accepted that another 200 units of resource are under a national park; but it is believed that production there will never be allowed.
  (a) What are price P and production X today (i.e., year 2001)?
  (b) Public policy is suddenly changed to allow production in the national park. What changes? Forecast P, X, and S five years from now (i.e., year 2006).
  (c) After five years of drilling, it is realized that the 200 units under the park was overestimated by a factor of 2. What changes? What do P and X become now (year 2006, immediately following the realization)?
  (d) Sketch the evolution of P, X, and S over time from year 2001 to 2010.

17 Costless competitive production of a sterile resource. Demand is

$$X = \frac{100}{P^{\frac{1}{3}}} \tag{1.172}$$

$S_0 = 1,000$, interest rate $r = .05$, and the ceiling price is infinite. You are considering selling production rights to all of this resource.
  (a) What price should you expect?
  (b) The buyer expects that there is a ceiling price $\bar{P} = 8$. What is the most this buyer will pay?
  (c) You have secret knowledge about a substitute product that your own R&D unit has developed. This substitute makes $\bar{P} = 6$. With the offer from the buyer above in hand, you should be happy. Quantify your happiness.

18 Derive Equation 1.128 for the time to peak when discovery rate is fast relative to production.

19 Simulate the discovery system described in the text, Section 1.3.1, under three scenarios: (a) fast discovery ($\rho = .1$, $\beta r = .05$); (b) moderate discovery ($\rho = .02$, $\beta r = .05$); and (c) slow discovery ($\rho = .005$, $\beta r = .05$). Use $S_0 = 100$ and $U_0 = 600$; assume the ceiling price $\bar{P}$ is infinite. Do the simulation results agree with the text discussion?

20 Derive the discovery solutions for S and U as in the text, but with "optimistic" production that anticipates future discovery:

$$X = \beta r(S + \alpha U) + \frac{a}{\bar{P}^\beta} \tag{1.173}$$

with $0 < \alpha < 1$. (a) Compare the results with the pessimistic solutions given in the text. (b) What happens when $\alpha > 1$? Is there a critical (large) value of α? Explain why or why not.

21 Redo the simulations in Problem 19, but with "optimistic" production:

$$X = \beta r \left(S + \frac{U}{2} \right) \tag{1.174}$$

as in Problem 20. Assume the ceiling price $\bar{P}$ is infinite. Compare these simulations to a) the analytic solutions from Problem 20; and (b) the pessimistic production simulations in Problem 19.

22 Equations 1.39 and 1.40 give the point beyond which rent decreases as S increases, for costless competitive production with linear demand.
  (a) Derive those equations.
  (b) Confirm the root given in the text.
  (c) Plot R versus S^* for several combinations of the fixed parameters $\bar{P}$, r, and b, and confirm that 1.39 and 1.40 are true.

23 Ten units of a nonrenewable resource are available under costless production. The market for it is described by the demand function

$$P = \sqrt{1 - X^2} \tag{1.175}$$

where P is the price and X is the production rate. There is no exploration; interest rate $r = 0.10$.
  (a) Confirm: $\epsilon = -X^2/P^2$.
  (b) Confirm: $Q \equiv P(1 + \epsilon) = 2P - 1/P$.
  (c) Confirm: Given Q, then $P = \frac{1}{4}\left(Q + \sqrt{Q^2 + 8}\right)$.
  (d) What are the limits on P? X? Q?
  (e) What is the Terminal Condition?
Find the initial price, the initial production rate, and the time to exhaustion under (f) competitive production; and (g) monopoly production. Do (f) and (g) by simulation.

24 Costless production of a sterile nonrenewable resource; $S_0 = 100$; $r = 6\%$; $\beta = .5$. It is observed that demand is growing linearly with time; in the coming year, it will grow by 4% and will continue linearly forever.
  (a) What is the production from now until exhaustion? Compute and plot X and P.
  (b) Compare with the no-growth-in-demand scenario, all other things being unchanged.

(c) Simulate (a) and (b).

(d) Compute the rent under (a) and (b).

25 Costless production of a sterile resource. Demand is

$$X = \frac{1}{\sqrt{P}} \tag{1.176}$$

Interest rate $r = .10$. Today, it is known that $P = 25$. There is no ceiling price.

(a) What are S and X today?

(b) What is the financial value of exclusive production rights to this resource?

(c) How long will it take before S is depleted to 25% of today's amount?

26 Continuation of Problem 25. A new conservation technology is about to be introduced. As a result, demand will shrink gradually over time:

$$X = \frac{e^{-\alpha t}}{\sqrt{P}} \tag{1.177}$$

with $\alpha = .06$ and $r = .10$. Note: Here you may still assume that $\frac{dP}{dt} = rP$. P_0 may change, however.

(a) What are P, X, and S as functions of time?

(b) What is today's price? Today's production rate? How have they changed from your Problem 25 answers? (In problem 25, $\alpha = 0$.)

(c) How long will it take before S is depleted to 25% of today's amount?

(d) What is the total value today of this new technology? Who is happy, customers, technologists, and/or resource owners?

27 Redo Problems 25 and 26; but add a price ceiling $\overline{P} = 15$. In addition to the questions there, evaluate the resource lifetime in each case.

28 Costless, competitive production of a sterile resource. Demand is

$$X = \frac{200}{\sqrt{P}} \tag{1.178}$$

There is no ceiling price; interest rate $r = .10$. There is exploration. But producers always act conservatively, forecasting zero discovery until it actually happens. Production one year ago was $X = 100$. Today, $X = 120$.

(a) What is the discovery rate?

(b) What are the total reserves S today?

(c) A company wants to buy exclusive rights to your resource. What will you sell for, today?

(d) The Supreme Court just ruled that 500 units of your S actually lie under a national park; your right to that amount is voided. Quantify your (un)happiness as an owner.

29 Costless production, no ceiling price, with constant-elasticity demand, shrinking exponentially:

$$X = \frac{e^{-\alpha t}}{P^\beta} \qquad (1.179)$$

What is the relation between the decay rate α and P_0? between α and total rent?

30 Costless production, constant elasticity $\beta = .5$. Parameters: $S_0 = 1,000$, $r = .05$, $a = 1$.

    (a) Compute and plot X, S, and P for the first 10 years of production.

    (b) At $t = 10$, an attractive substitute for the resource becomes instantly available. As a result, demand begins to shrink exponentially, at the rate $g = -0.20$ per year. What changes in X and P will occur instantly?

    (c) Plot the time history of X, S, and P from $t = 0$ (as in (a) above) and continuing through the change in demand, out to $t = 30$.

31 You own 6,000 units of a sterile resource; production is costless, and there is no discovery. Demand is

$$X = \frac{7}{\sqrt{P}} \qquad (1.180)$$

Interest rate $r = .05$ and there is no price ceiling.

    (a) What is today's price?

    (b) How much resource is left after 10 years?

    (c) At $t = 10$ years, you decide to sell all rights to future production. How much money will you sell out for?

    (d) Engineers in another firm have worked secretly on a new use for your resource. You are ignorant of this. This new technology is "ready to go" at $t = 10$. Once announced, it is expected that demand will instantly expand to

$$X = \frac{9}{\sqrt{P}} \qquad (1.181)$$

    Assuming they first buy the production rights from you as in (c) above, quantify their happiness.

    (e) The technical announcement about the new use in (d) is made, at $t = 10$. What immediate adjustments will occur in price, supply, and production rate? With this in hand, sketch the time history of these quantities from today $(t = 0)$ through $t = 15$.

32 Costless production of a sterile resource, with demand

$$X = \frac{7}{\sqrt{P}} \qquad (1.182)$$

Interest rate $r = .05$, and there is no price ceiling. You own 6,000 units of the resource today.

(a) What is today's price?

(b) What is today's production rate?

You read the following in the evening news: "*The government announced today that, effective in eight years, production of this resource will be illegal due to unacceptable health risks.*"

(c) What changes should you make?

(d) What is the new price, effective tomorrow morning?

(e) What is the new production rate, effective tomorrow morning?

(f) You will sue the government for damages. What is the dollar value of your suit?

33 Costless production of a sterile resource, with demand

$$X = \sqrt{1 - P} \tag{1.183}$$

Interest rate $r = .05$. It is known that in exactly four years, this resource will be exhausted.

(a) What is the production rate and price, both now and in four years, under *competitive* production? (Your answer should consist of four numbers.)

(b) Same, under *monopoly* production.

34 Compute the present worth of the consumers' surplus, $\int CS \cdot e^{-rt} dt$, as given in Equation 1.27, assuming $P(t) = P_0 e^{rt}$.

35 For linear demand given by

$$P = \overline{P} - bX \tag{1.184}$$

(a) Find an expression for consumers' surplus as a function of P.

(b) Find an expression for the present worth of consumers' surplus, $\int CS \cdot e^{-rt} dt$ assuming $P(t) = P_0 e^{rt}$.

2 Biomass

Here we introduce the case where the resource itself reproduces, occupying Quadrants 3 and 4 simultaneously. We begin with a single resource stock: biomass B. It is necessary to account for its growth rate and for the harvest. Steady states are possible when these two are in balance, Quadrant 3; but the descent into Quadrant 4, "resource mining," occurs when harvest continuously exceeds growth, leading to extinction: no S, therefore no Q. This would be the analog of resource exhaustion for the sterile resource. It is irreversible.

These resources are local in that they live within a regional ecosystem with finite carrying capacity. The market for the harvest, however, is presumed exogenous, dependent on many things other than this particular resource. The application to fishery management is used throughout to fix ideas.

The base case here uses the logistic function for the growth rate. As the resource (e.g., fish) is fugitive, harvesting requires effort (fishers) as well as fish availability. The formulation needs to add fishing effort E at a fundamental level. The economic interaction of growth and harvesting is commonly referred to as "bioeconomic." From a sustainability perspective, there needs to be attention to (a) avoiding the "mining" phenomenon associated with Quadrant 4 extinction; (b) regulating the effort directed at the resource harvesting; and (c) respecting the conditions required for maintenance of the reproducing stock (the ecological carrying capacity).

2.1 GROWTH AND HARVESTING

In the case of sterile resources, we have only one consideration: the rate of its exhaustion and the time frame of complete exhaustion. Exhaustion can be "physical" exhaustion, as in the case of costless production. More realistically, "economic" exhaustion would indicate that the resource can no longer be produced economically – the cost of production exceeds its value. "Political" exhaustion occurs when the resource cannot be produced legally.

By contrast, living systems present the possibility of sustained resource usage, indefinitely. We will therefore be concerned with the possibility of steady states, reflecting a balance between nature and the economy, and the stability of those states. We will start with a simple situation where the living system is characterized by a single descriptor, the biomass B.

The basic natural dynamic for this system is

$$\frac{dB}{dt} = G - H \tag{2.1}$$

with

- $B(t) =$ amount of biomass
- $G(t) =$ biological growth rate
- $H(t) =$ harvesting rate

To close this system, we need relations for both growth and harvesting rates.

2.1.1 Growth

A candidate growth rate function is the logistic function

$$G(B) = gB\left(1 - \frac{B}{K}\right) \tag{2.2}$$

with K the carrying capacity of the system. This growth rate depends on B only. There is positive growth at all positive levels of B, up to the carrying capacity, beyond which G is always negative.

At low B, $G \simeq gB$ and in the absence of harvesting, $dB/dt = gB$, and we expect exponential growth, $B = B_0 \exp(gt)$. This would be a frontier situation, characteristic of, for example, an invasive species in its early history.

At intermediate B, G increases with B, reaching a peak at $B^* = K/2$. Higher B leads to decline in G. The *proportional* rate of growth G/B declines monotonically with B:

$$\frac{G}{B} = g\left(1 - \frac{B}{K}\right) \tag{2.3}$$

This might occur due to crowding, habitat restrictions, food limitation, or attraction of predators.

As $B \to K$, growth shuts off. The simple substitution $\epsilon = K - B$ gives, in the absence of harvesting and near carrying capacity, $d\epsilon/dt = -g\epsilon$. At carrying capacity, B is stable. Departures from K will decay exponentially as $\exp(-gt)$.

Logistic Growth

Figure 2.1 illustrates the logistic growth of B over time and the attendant G history, beginning at very low B and approaching K in the absence of harvesting. The steepest growth occurs midway in the trajectory, at intermediate B.

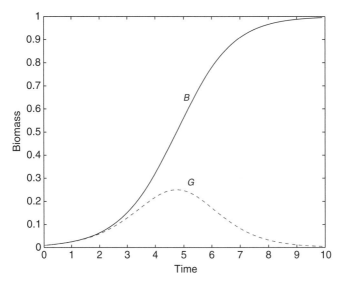

Figure 2.1. Time evolution of B and growth rate G in the absence of harvesting. Logistic growth $G = B(1 - B)$.

The governing equation is

$$\frac{dB}{dt} = gB\left(1 - \frac{B}{K}\right) \tag{2.4}$$

and its solution is

$$B(t) = \frac{K}{1 + Ae^{-gt}} \tag{2.5}$$

$$A = \frac{K - B_0}{B_0} \tag{2.6}$$

where B_0 is the initial condition at time $t = 0$. If initial conditions are small, then A is big. Peak growth occurs at $B = K/2$, at time T_p:

$$Ae^{-gT_p} = 1 \tag{2.7}$$

and therefore

$$gT_p = \ln(A) = \ln\left(\frac{K}{B_0} - 1\right) \tag{2.8}$$

In the limit of small B_0/K,

$$gT_p \simeq \ln\left(\frac{K}{B_0}\right) \tag{2.9}$$

As an example, suppose an invasive species with $g = 0.1/\text{yr}$ is observed at 1% of carrying capacity. The waiting time to peak G would be $T_p = 46$ years.

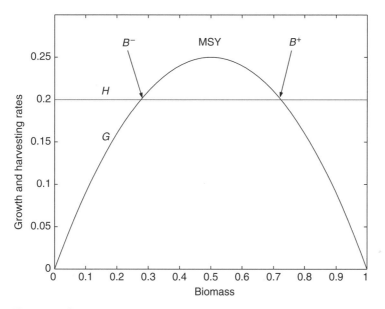

Figure 2.2. Growth and harvesting rates. The case of constant harvest is shown, with logistic growth function $G = B(1 - B)$. There are two intersections where $G = H$; the one on the left, B^-, is unstable.

Steady State

For steady state, we need $G = H$. The highest harvest possible occurs at $B = B^*$, with $H^* = G^* = gK/4$. This is the maximum sustainable yield (MSY). For any fixed value of harvest $H < H^*$, there are two equilibria, B^+ and B^-, symmetrically situated about B^* (Figure 2.2). If harvesting were arranged to be constant, then the right-most of this pair, B^+, would be stable to small perturbations in B; while the left-most B^- would be unstable. Negative perturbations about B^- would result in extinction; positive perturbations would result in growth toward the stable equilibrium at B^+. Harvesting in excess of H^* also results in extinction, as no growth could ever keep up. Thus, we have two recipes for extinction: Operate at low biomass, $B < B^-$, *or* harvest above MSY, $H > H^*$. And there are two conditions for a stable, sustainable harvest: Harvest below MSY, *and* avoid the possibility of large negative disturbances to B, such that B falls below B^-.

2.1.2 Harvest

In considering the harvest rate, we need the concept of harvesters' effort: the number of jobs, machines, etc. involved in active harvesting and their relative employment, activity, or utilization (e.g., number of days per year spent harvesting). We will lump all these factors into a single effort variable E.

The harvest depends on the effort and the biomass. A simple relation is

$$H(B, E) = hEB \tag{2.10}$$

wherein h represents the harvesting technology. Increases in effort, biomass, or technology increase the harvest, linearly in this case. Absence of any of these factors guarantees zero harvest.

With this closure, we can describe the steady state in which $H = G$:

$$hEB = gB\left(1 - \frac{B}{K}\right) \tag{2.11}$$

and thus,

$$E = \frac{g}{h}\left(1 - \frac{B}{K}\right) \tag{2.12}$$

or equivalently,

$$\frac{hE}{g} + \frac{B}{K} = 1 \tag{2.13}$$

So for this fishery in steady state, E is linear in B. We may operate at any combination of B and E (fish and fishers) on this line; at one extreme, $(E, B) = (0, K)$, and we have the natural carrying capacity with no effort and no harvest; while at the other extreme, $(E, B) = \left(\frac{g}{h}, 0\right)$, we have a system at infinitesimal biomass, essentially no harvest, and much effort devoted to keeping it there. In the middle, we have the MSY point $(E, B) = \left(\frac{g}{2h}, K/2\right)$ with $H = H^*$. If effort were able to be controlled, we could choose among these equilibria or any other (E, B) pairs along the line represented by Equation 2.12.

Figures 2.3 and 2.4 illustrate logistic and depensatory growth, with harvesting as in Equation 2.10.

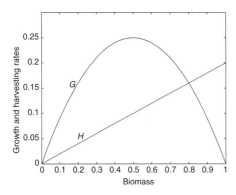

Figure 2.3. Gowth and harvesting rates with $H = hEB$, for a representative value $hE = .2$. The logistic growth function is compensatory at low B; the unstable intersection is absent.

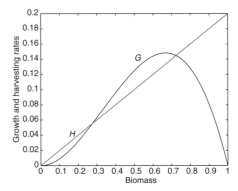

Figure 2.4. Growth and harvesting rates. $H = hEB$ as in Figure 2.3, but $G = B^2(1 - B)$. This growth curve is depensatory at low B, with an unstable descent to extinction in that low B range.

2.1.3 Rent

Next, introduce rent as the net profit resulting from the harvest. We will take the sales price for the harvest, p, to be constant and the wage or cost of effort, c, also as constant.[1] So, rent R is given by

$$\pi = pH - cE \tag{2.14}$$

In the steady state, $H = G$ and E is given by Equation 2.12, so

$$\pi(B) = pgB\left(1 - \frac{B}{K}\right) - c\frac{g}{h}\left(1 - \frac{B}{K}\right) \tag{2.15}$$

Now we can characterize three interesting steady-state points above by the four descriptors B, H, E, and R (fish, food, effort, and money).

Point 0: (resource extinction)

$$B = 0$$
$$H = 0$$
$$E = g/h \tag{2.16}$$
$$\pi = -cg/h$$

Point MSY: (maximum sustainable yield)

$$B = K/2$$
$$H = gK/4 = H^* $$
$$E = g/2h \tag{2.17}$$
$$\pi = pgK/4 - cg/2h$$

Point K: (carrying capacity)

$$B = K$$
$$H = 0$$
$$E = 0 \tag{2.18}$$
$$\pi = 0$$

It is interesting to notice that the MSY point may produce either positive or negative rent; it is not characterized by an economic criterion, but rather by a biological one. Point K is also an exclusively biological one, with no harvesting effort and no rent.

---

[1] Here we assume large external markets in food (p) and effort (c). This resource is on the margin of a large economy, which it does not affect; p and c are constants.

Point 0, however, requires maximal effort to maintain the level $B = 0$. Many other equilibria are possible, between points $B = 0$ and $B = K$, with an implied steady-state trade-off between E and B.

> *Aside*: It is interesting to reexpress rent in terms of B and H. When $H = hEB$, we always have
>
> $$E = \frac{H}{hB} \tag{2.19}$$
>
> and therefore,
>
> $$\pi = pH - cE \equiv \left(p - \frac{c}{hB}\right) H \tag{2.20}$$
>
> (This does not assume a steady harvest.) Immediately, we have $\pi = \pi(H, B)$, and the "cost of harvest" is c/hB – that is, dependent on B. This is a "stock effect." We also identify $B = c/hp$ as the condition of zero rent; this parameter B_0 is used extensively below:
>
> $$B_0 \equiv \frac{c}{hp} \tag{2.21}$$
>
> The general rent formula is as above, $\pi = \pi(H, B)$. In the steady state, with $G = H$,
>
> $$\pi_s = \pi(G, B) \tag{2.22}$$
>
> $$= \left(p - \frac{c}{hB}\right) G \tag{2.23}$$
>
> $$\frac{d\pi_s}{dB} = \left(p - \frac{c}{hB}\right) \frac{\partial G}{\partial B} + \left(\frac{c}{hB^2}\right) G \tag{2.24}$$
>
> This formulation of π is a useful alternative to that formulated explicitly in terms of effort, $\pi = pH - cE$.

Which equilibrium is likely to occur? That depends on the conditions under which the fishery is operated and the criterion on which that is based.

2.2 ECONOMIC DECISION RULES

Here we will assume that harvesting can be made profitable over some range of B, as in Figure 2.5. The opposite is easy to envision – for example, when c is very large and rent is always negative.[2] Under these conditions, the resource is *economically extinct*; economic actors would abandon all harvesting, and the steady state would be at the carrying capacity, point K, with no economic effort, harvest, or rent. A steady harvest does not pay.

Assuming harvesting is profitable over some range, we have two interesting steady states.

---

[2] This occurs in this fishery when $c/ph > K$, that is, $B_0 > K$.

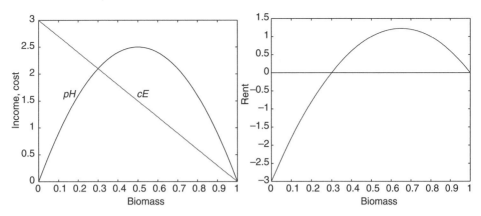

Figure 2.5. Income pH and expenses cE (*left*) and rent π (*right*), as a function of B, in steady state for logistic growth $G = gB(1 - B/K)$ and $H = hEB$. Parameters: $(g, h, K) = 1$; $p = 10$; $c = 3$.

2.2.1 Free-Access Equilibrium

In a *laissez-faire* system, with free access to the fishery, we hypothesize that new effort (fishers) will enter the business as long as the value of the harvest exceeds the cost of its production – essentially, as long as $\pi > 0$. (Recall that "cost" includes the cost of all inputs to production, *including reasonable/necessary return on investment* in equipment; positive rent implies a windfall situation where all of these costs are paid, including all returns on captital invested or borrowed; and there is still income left over.) The condition of vanishing steady rent (Equation 2.15) characterizes this equilibrium:

$$\pi(B) = pgB\left(1 - \frac{B}{K}\right) - c\frac{g}{h}\left(1 - \frac{B}{K}\right) = 0 \tag{2.25}$$

Setting $\pi(B) = 0$ gives the equilibrium at $B = \frac{c}{ph} \equiv B_0$; we identify this equilibrium value as B_0 for convenience. This is the "open-access," or "free-entry," point. Because positive rent is possible at lower effort, this point is sometimes characterized as the "rent dissipation" point, arrived at by increasing E freely until π is driven to zero. Key quantities at this point are

Point R_0: (free entry)

$$B = B_0 = \frac{c}{ph}$$

$$H = gB_0\left(1 - \frac{B_0}{K}\right)$$

$$E = E_0 \equiv \frac{g}{h}\left(1 - \frac{B_0}{K}\right) \tag{2.26}$$

$$\pi = 0$$

The point R_0 is stable under these conditions. Effort may not increase without encountering negative rent, and vice versa.

There is another point where rent vanishes, at $B = K$. It is economically unstable under open access; $E = 0$ there, but positive rent at $E > 0$ would encourage entry of effort, such that the stable equilibrium R_0 is approached, with $B < K$ and $E > 0$.[3]

2.2.2   Controlled-Access Equilibrium

In a *monopoly-operated* system, effort may be controlled. It is visually apparent that more rent may be earned by reducing effort relative to the open-system equilibrium (B_0). The controlled-access equilibrium is defined by the condition of rent maximation. Taking the derivative of Equation 2.15 with respect to B gives us the extremum condition:

$$\frac{d\pi(B)}{dB} = pg - 2\frac{pg}{K}B + \frac{cg}{hK} = 0 \qquad (2.27)$$

Solving for B gives the equilibrium:

Point R^*: (maximum rent, controlled access)

$$B = \frac{K + B_0}{2}$$

$$H = gK\left(1 - [B_0/K]^2\right)/4$$

$$E = E_0/2 = \frac{1}{2}\frac{g}{h}\left(1 - \frac{B_0}{K}\right) \qquad (2.28)$$

$$\pi = \pi_{MSY} + (pg/4K)B_0^2$$

This point requires monopoly control of effort.[4] Compared with the open-access case, we have less effort, more (positive) rent, and higher B. A standard characterization is that there is less work being done, more money made, and a higher biomass. The effect on harvest H is ambiguous and depends on the parameters.

Table 2.1 summarizes the five steady states discussed so far.

Table 2.1. Steady-state operating options for $H = hEB$; $G = gB(1 - B/K)$. $B_0 \equiv c/ph$ is the open-access (free-entry) equilibrium. When $B_0 > K$, the resource is economically extinct.

		B	H	E	π
0	Extinction	0.	0.	g/h	$-cg/h$
MSY	Maximum harvest	$K/2$	$gK/4 = H^*$	$g/2h$	$pgK/4 - cg/2h$
K	Carrying capacity	K	0.	0.	0.
R_0	Free entry	$B_0 \equiv c/ph$	$gB_0(1 - B_0/K)$	$E_0 \equiv g(1 - B_0/K)/h$	0.
R^*	Maximum rent	$(K + B_0)/2$	$gK\left(1 - [B_0/K]^2\right)/4$	$E_0/2$	$\pi_{MSY} + (pg/4K)B_0^2$

---

[3]   As above, this assumes $B_0 \equiv c/ph < K$; otherwise, the resource is economically extinct.

[4]   Notice that for the costless case, point R^* is the same as point MSY.

Control of harvesting is often conceived as one or more measures aimed at effort control, harvest control, technology control, market control, or political control. All are normally achieved by a permitting/inspection process. There are several generic forms:

- Effort at harvesting
  - Days at sea
  - Vessel licensing
- Harvest control
  - Count landed harvest
  - Inspection at sea
  - Quotas
- Technology control
  - Information technology
  - Vesssel size, speed
  - Method of capture/storage
- Market control
  - Tax the product
  - Tax the landed harvest
  - Tax the effort
- Political control
  - Penalties for violation: civil, criminal, economic

2.3 EFFORT DYNAMICS

It is easy to extend a formal description of effort adjustment. The discussion above describes this in terms of rent: When $\pi > 0$, effort in the open-access system increases; and when $\pi < 0$, effort diminishes. We formalize this with a first-order rate $\nu\pi$. Together with the system described so far, we have the dynamic system

$$\frac{dB}{dt} = G - H \tag{2.29}$$

$$H = hEB \tag{2.30}$$

$$G = gB\left(1 - \frac{B}{K}\right) \tag{2.31}$$

$$\pi = pH - cE \tag{2.32}$$

$$\frac{dE}{dt} = \nu\pi \tag{2.33}$$

There are two dynamic state variables (B, E) and three constitutive relations defining the auxiliary variables (H, G, π). Fixed parameters include ν, h, g, K, p, and c. As a second-order nonlinear system, we have the potential for extravagant behavior, and the parameter ν remains to be explored.

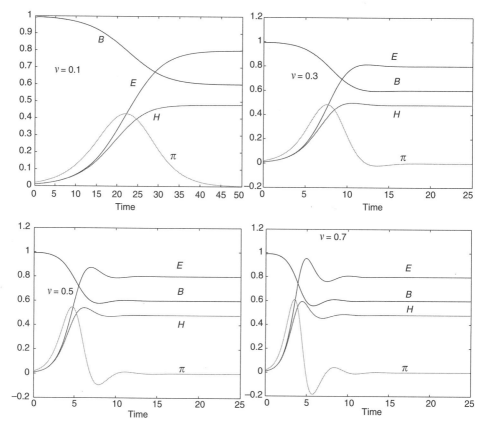

Figure 2.6. Dynamics of the free-access solution as a function of v, as indicated. Logistic growth $G = gB(1 - B/K)$ and $H = hEB$. Parameters: $(g, h, K) = (2, 1, 1)$; $(p, c) = (5, 3)$. Euler (explicit forward) integration, $\Delta t = 0.1$. Equilibrium values are $(B, E, H, \pi) = (0.6, 0.8, 0.48, 0.0)$.

The program **fish1_4.m** (**Fish1.4.xls**) illustrates this; results appear in Figure 2.6. Equilibrium values are $B = 0.6$, $E = 0.8$, $H = 0.48$, and $\pi = 0.0$. There is an orderly approach to equilibrium at low v; during this approach, rent rises, peaks, and returns toward zero as the equilibrium is approached. Exploitation during this period is lucrative. Increasing v to 0.3 speeds up the process but otherwise adds little to the dynamic. $v = 0.5$ and 0.7 continue this trend; in addition, an overshoot is introduced, following which rent becomes negative, E reduces (an employment layoff), and H recovers, all in an orderly approach to the same equilibrium.

Figure 2.7 presents the high v extension of this case. The equilibrium solutions are unchanged, yet all hope for monotone solutions is gone in these rapid-response-to-rent scenarios. The overshoot trend identified above at low v is here amplified, and (ultimately) wild oscillations are evident in B, E, π (i.e., economy, employment, and ecology). A complex periodicity of about two years is apparent. Amplitude increases with v; the damping rate decreases with v. High v clearly is producing boom-or-bust cycling in this system.

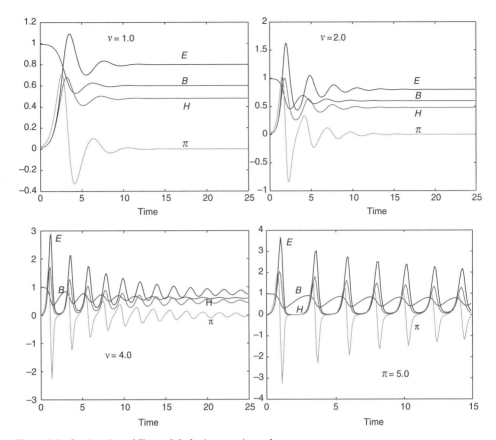

Figure 2.7. Continuation of Figure 2.6, for larger values of *v*.

Program **Fish1.4-RAND.xls** adds independent stochastic disturbances to this dynamic, through G and dE/dt. These are autocorrelated as in the Appendix. Program **Fish1.4-Cubic.xls** uses cubic growth function, with depensation exaggerated by the incorporation of a minimum viable biomass $\underline{B}$:

$$G(B) = gB(B - \underline{B})(1 - B/K) \tag{2.34}$$

Either depensatory growth, or the entry of stochastic disturbances, can exaggerate the negative effects of rapid effort dynamics (high v).

2.4 INTERTEMPORAL DECISIONS: THE INFLUENCE OF *r*

In previous analyses, we ignored the effect of r, the growth rate of money invested in a productive economy. We have concentrated solely on the sustainable harvest, correct for $r = 0$. Here we will reinstate the effect of $r > 0$.

We imagine a situation where a harvester discovers a resource in the unexploited state $B = K$. Faced with the option to establish a steady-state harvest, B must be reduced to some level $B < K$. That initial harvest, assumed instantaneous for the moment, would be profitable and the proceeds invested at interest rate r.

2.4.1 Costless Harvesting

The sale of the initial harvest is $p(K - B)$, which is invested at interest rate r. The harvest at B is then sustained, and its sale value is $pG(B)$. Annual income π is thus the sum of the sale of the sustainable harvest plus the investment earnings from the initial harvest:

$$\pi = pG(B) + rp(K - B) \tag{2.35}$$

Its maximum is found by differentiating

$$\frac{d\pi}{dB} = p\frac{dG}{dB} - rp = 0 \tag{2.36}$$

and the optimal point is at

$$\frac{dG}{dB} = r \tag{2.37}$$

For the logistic growth curve

$$G(B) = gB\left(1 - \frac{B}{K}\right) \tag{2.38}$$

we have

$$\frac{dG}{dB} = g\left(1 - \frac{2B}{K}\right) \tag{2.39}$$

Figure 2.8 illustrates this balance. The steady B will always be below the MSY point, as dG/dB is negative above that. The maximum growth rate is g; when $r > g$, extinction is the "rational" solution. Otherwise, the sustainable solution is between extinction and the MSY point.

It is clear that the scenario given initially can be relaxed; starting from *any* initial B, we arrive by the same reasoning at the desired equilibrium: balancing the annual yield of the initial harvest against the annual yeild of steady harvesting.

The consequence of this is striking: Such a "rational economic actor" with guaranteed and exclusive access to the resource would always find the steady equilibrium B and H on the rising limb of the growth curve, *below* the MSY point.[5] For slow-growing resources, where $dG/dB < r$ irrespective of abundance B, such an approach would result in extinction; the resource growth cannot equal financial investments. Money grows faster than the resource under all conditions!

---

[5] Under costless harvesting, point MSY is also the maximum rent equilibrium R^* identified above.

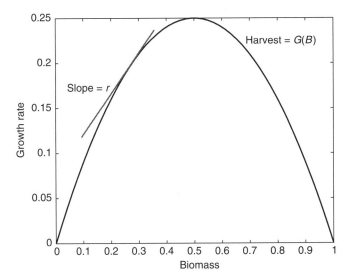

Figure 2.8. Growth curve indicating the intertemporal equilibrium where $dG/dB = r$, for costless harvesting. The equilibrium is at $B \simeq .28$ in this illustration.

It would appear from this analysis that a strictly financial criterion for a "renewable" resource is $dG/dB > r$; resources with slower growth would be exploited as if they were sterile and exhaustible under free access.

Notice that here we have no explanation for r, unlike the biology that creates G. A fuller examination of this balance would need to develop the relationship between the growth rate of money invested and the existence and growth of natural resources.

Notice, too, that we have now developed two different criteria for *economic extinction*:

- When there is B but it is too costly to harvest, then $B \to K$.
- When G is too slow, it is attractive to harvest all of B and invest it in the economy at growth rate r, *presumed sustainable*.

These are dramatically different situations!

2.4.2 Costly Case

The same exercise with costly harvesting can be done. Consider operating at the harvest rate $H = hBE$, with sustainable annual rent $\pi_s = \pi_s(B)$, as in Figure 2.5. We contemplate an *instantaneous surge in effort* ΔE for a small period of time Δt (Figure 2.9); with a related instantaneous surge in harvest ΔH over the same small Δt:

$$\Delta H = hB\Delta E \tag{2.40}$$

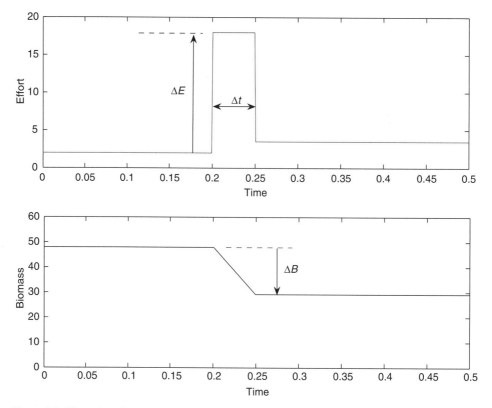

Figure 2.9. Illustration of an instantaneous surge in *E* and the resulting change in *B*. Both levels of B are to be harvested sustainably.

The instant, one-time rent quantum would be

$$\Delta \pi_i = (p\Delta H - c\Delta E)\Delta t \tag{2.41}$$

$$= \left(p - \frac{c}{hB}\right)\Delta H \Delta t \tag{2.42}$$

As a result of this change, all future harvests will be based on a decreased *B*:

$$\Delta B = -\Delta H \Delta t \tag{2.43}$$

and the sustainable rent $\pi_s(B)$ would change by the amount

$$\Delta \pi_s = \frac{d\pi_s}{dB}\Delta B \tag{2.44}$$

The two effects from the instantaneous change are: $\Delta \pi_i$, the instant, one-time rent quantum in the bank (due to the instantaneous ΔE and ΔH); and the changed sustainable rent $\Delta \pi_s$ (due to the permanent change in ΔB). On an annual basis, the net change is

$$\Delta\pi = \Delta\pi_s + r\Delta\pi_i \tag{2.45}$$

$$= \left[\frac{d\pi_s}{dB} + r\left(-p + \frac{c}{hB}\right)\right]\Delta B \tag{2.46}$$

$$= \left[\frac{d\pi_s}{dB} - rp\left(1 - \frac{B_0}{B}\right)\right]\Delta B \tag{2.47}$$

(Earlier we introduced the open-access equilibrium $B_0 \equiv c/ph$.) Indifference about this trade-off must characterize the equilibrium point; $\Delta\pi = 0$:

$$\frac{d\pi_s}{dB} = rp\left(1 - \frac{B_0}{B}\right) \tag{2.48}$$

When costs are zero, we recover the costless case above (Equation 2.37), with $\pi_s = pG$. This result generalizes that. Clearly, the controlled access point has

$$\frac{d\pi_s}{dB} = 0 \tag{2.49}$$

and assuming profitable harvesting, then here we have

$$\frac{d\pi_s}{dB} > 0 \tag{2.50}$$

and we can see that we are moving to lower B in Figure 2.5. The intertemporal consideration has required sustainable operation at a lower B than we had before: It is profitable to cut back on B once, invest the proceeds in the economy, and harvest the interest plus the remaining $G(B)$ sustainably thereafter. When $r = 0$, we recover the previous equilibrium R^* (the rent-maximizing controlled access point).

Base Case: Logistic Growth

For the base case

$$G = gB\left(1 - \frac{B}{K}\right) \tag{2.51}$$

$$H = hEB \tag{2.52}$$

we have from Equations 2.15 and 2.27

$$\pi(B) = pgB\left(1 - \frac{B}{K}\right) - c\frac{g}{h}\left(1 - \frac{B}{K}\right) \tag{2.53}$$

$$\frac{d\pi(B)}{dB} = pg - 2\frac{pg}{K}B + \frac{cg}{hK} \tag{2.54}$$

Assembling Equation 2.48, we obtain

$$1 - 2\frac{B}{K} + \frac{B_0}{K} = \frac{r}{g}\left(1 - \frac{B_0}{B}\right) \tag{2.55}$$

A little algebra gives the quadratic equation

$$\beta^2 - \beta \frac{(1 + \beta_0 - \rho)}{2} - \frac{\rho \beta_0}{2} = 0 \tag{2.56}$$

where we have introduced the normalized variables

$$\beta \equiv B/K \tag{2.57}$$

$$\rho \equiv r/g \tag{2.58}$$

$$\beta_0 \equiv B_0/K \equiv c/phK \tag{2.59}$$

The roots of this are[6]

$$\beta = \left[\frac{1 + \beta_0 - \rho}{4}\right] \pm \sqrt{\left[\frac{1 + \beta_0 - \rho}{4}\right]^2 + \frac{\rho \beta_0}{2}} \tag{2.60}$$

This result reproduces earlier results:

- In the costless case ($\beta_0 = 0$), $\beta = \frac{1-\rho}{2}$
- In the no-interest case ($\rho = 0$), $\beta = \frac{1+\beta_0}{2}$

In the interesting intermediate cases, $\beta_0 < \beta < \frac{1+\beta_0}{2}$. In the limit of vanishing ρ, we get the closed-access equilibrium case identified above. As ρ increases, β is reduced

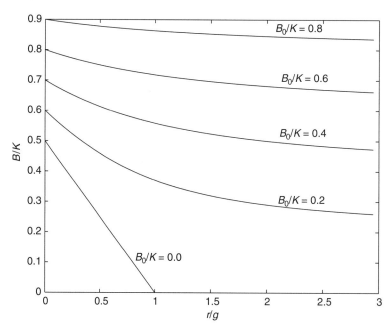

Figure 2.10. Equilibrium $\beta \equiv B/K$ versus $\rho \equiv r/g$ from Equation 2.60. This is the intemporal, rent-maximizing equilibrium for the base case (logistic) fishery. $B_0 \equiv c/ph$ is the open-access equilibrium.

---

[6] The positive option is correct here; the negative option is useless, returning $\beta < 0$.

toward its open-access counterpart β_0. Programs **Fish1.6.xls** and **Fish1_6.m** simulate this. Figure 2.10 illustrates this.

2.4.3   Costly Case: A General Expression

An alternative, and equivalent, view of the intertemporal case is useful and common. Here we note that with $H = hEB$, we always have

$$E = \frac{H}{hB} \tag{2.61}$$

and thus the sustainable rent is

$$\pi = pH - cE \tag{2.62}$$

$$= \left[p - \frac{c}{hB} \right] H \tag{2.63}$$

$$= \pi(H, B) \tag{2.64}$$

The cost term c/hB has a dependence on the resource "stock" B as well as the harvest H. In this formulation, the cost of harvesting increases as the biomass is reduced.[7] The partial derivatives are

$$\frac{\partial \pi}{\partial B} = \frac{cH}{hB^2} > 0 \tag{2.65}$$

$$\frac{\partial \pi}{\partial H} = p - \frac{c}{hB} \tag{2.66}$$

More general harvesting rate functions $H(E, B)$ are possible and the form

$$\pi = \pi(H, B) \tag{2.67}$$

is quite general. In these general terms, the intertemporal trade-off is as before:

$$r \frac{\partial \pi}{\partial H} \Delta H \Delta t + \Delta \pi = 0 \tag{2.68}$$

with the first term coming from the one-time harvest and the second term coming from the reduced sustainable harvest. Expanding the second term, we have both the harvest and stock effects:

$$\Delta \pi = \frac{\partial \pi}{\partial B} \Delta B + \frac{\partial \pi}{\partial H} \frac{dH}{dB} \Delta B \tag{2.69}$$

$$= \frac{\partial \pi}{\partial B} \Delta B + \frac{\partial \pi}{\partial H} \frac{dG}{dB} \Delta B \tag{2.70}$$

(We assume the steady balance $H = G$.) Adding the fact $\Delta B = -\Delta H \Delta t$, the result is

$$r \frac{\partial \pi}{\partial H} = \frac{\partial \pi}{\partial B} + \frac{\partial \pi}{\partial H} \frac{dG}{dB} \tag{2.71}$$

---

[7] The "stock effect" on cost of production was seen in the nonrenewable case earlier, $C(S)$ in Chapter 1.

and a little rearrangement gives

$$\frac{dG}{dB} = r - \left[\frac{\partial\pi/\partial B}{\partial\pi/\partial H}\right] \tag{2.72}$$

This relation is quite general; it does not depend on the specific closures for $G(B)$ and $H(E, B)$ introduced above. It recovers the simple costless case, Equation 2.37, when c and therefore $\partial\pi/\partial B$ vanishes. Otherwise, $\partial\pi/\partial B > 0$ and the equilibrium moves toward smaller dG/dB, higher B.

When $r = 0$, this point recovers the equilibrium R^* (the peak sustainable rent for controlled access), as there is no value in financial investment and the trade-off is reduced to $d\pi/dB = 0$.

Equation 2.72 is referred to by several authors as the "fundamental equation of renewable resources" (e.g., Conrad (1999), Equations 1.16 and 3.5).

From the three demonstrations here, we can see that the general case puts the optimal intertemporal trade-off in between the simple closed-access optimum R^* (valid when $r = 0$) and the costless limit $dG/dB = r$.

2.5 TECHNOLOGY

The technology of harvesting has been assumed constant up to now. It clearly figures in our formulation of harvest rate, via the parameter h:

$$H = hEB \tag{2.73}$$

Advances in technology plainly amplify effort and make possible harvesting at lower biomass or effort or both. On what do h and dh/dt depend? A realistic treatment of technology is needed for a full theory of natural resource dynamics; here we simply speculate on the form it might take.

We denote by $I(t)$ the innovation rate:

$$\frac{dh}{dt} = I \tag{2.74}$$

Necessity is the mother of invention; and in a crude way, the purpose of entrepreneurship is to create rent where currently there is none. So we suppose that π drives innovation. That might take a form such as

$$I = I_0 e^{-\beta\pi} \tag{2.75}$$

At low rent, there is incentive to innovate; but as rent rises, there is less urgency. (Recall this is a rate of innovation.) Naturally, negative innovation and/or negative rent is not meaningful here, so we limit ourselves to the first quadrant of this plot (Figure 2.11). Implied is that $I = 0$ for $\pi \le 0$.

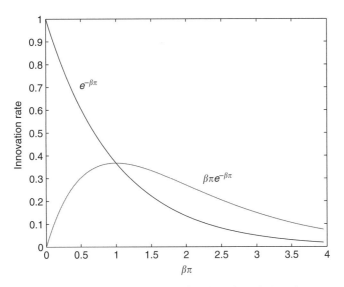

Figure 2.11. Candidate innovation rate functions. $I_0 = 1$; $\beta = 1$.

We can do better for low π. The abrupt termination there would be better modeled with a gentler approach to the origin. A candidate with the correct shape is

$$I = I_0 \beta \pi e^{-\beta \pi} \tag{2.76}$$

This form is also plotted in Figure 2.11. It contains the idea that if π is small, there is little reinvestment in harvesting technology. As rising π attracts innovation investment, whether from within the harvesting industry or from without. But it is self-extinguishing at very high π as in the previous form.

Finally, there is the possibility of technical saturation: a ceiling $\overline{h}$ that limits I:

$$I = I_0 \beta \pi e^{-\beta \pi} \left(\frac{\overline{h} - h}{\overline{h}} \right) \tag{2.77}$$

or a similar effect without the absolute ceiling:

$$I = I_0 \beta \pi e^{-\beta \pi} \left(\frac{\overline{h}}{h} \right) \tag{2.78}$$

The ultimate form suggested here has three parameters $(I_0, \beta, \overline{h})$ and two state variables (h, π):

$$\frac{dh}{dt} = I_0 \beta \pi e^{-\beta \pi} \left(\frac{\overline{h} - h}{\overline{h}} \right) \tag{2.79}$$

Program **Fish1.5.xls** simulates this. The technology grows during exploitation. Technology growth amplifies human effort, frees it to work on other things, and compensates for scarcity. But ultimately, we must confront its effect on the growing resource B. At constant E, for example, we find that increasing h eventually causes resource extinction in a free-entry system. Effectively, unbounded h brings us to the costless harvesting limit that, in a free entry system, drives $B \to 0$ (Figure 2.12). Some form of effort and/or technology control is needed on the part of the owner if this scenario is to be avoided.

The dynamics of this system are illustrated in Figure 2.13 for a case studied earlier (Figure 2.6). Clearly, the addition of innovation here has resulted in enhanced cycling of the fishery and a general trajectory toward lower biomass.

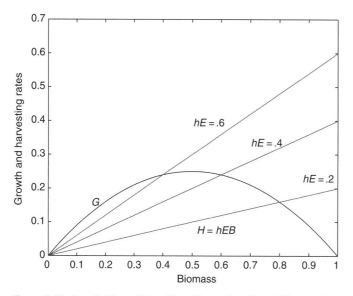

Figure 2.12. Growth ($G = gB[1 - B]$) and harvesting ($H = hEB$) rates. Increasing technology h moves the equilibrium toward the left (lower B) unless compensated for by decreasing effort E. Extinction is reached when $hE > g$ for these rates.

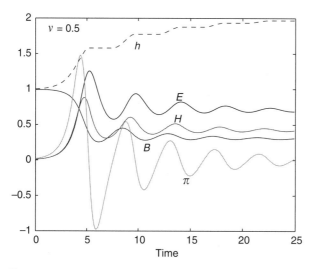

Figure 2.13. Dynamic adjustment as in Figure 2.6 for $v = 0.5$, but with innovation parameters $(I_0, \beta, \bar{h}) = (1, 0.5, 5)$ and initial $h = 1$.

Other forms of technological saturation functions $S(h)$ are useful. Above, we used the form

$$S(h) = 1 - \frac{h}{\bar{h}} \tag{2.80}$$

This has the anomaly that exceeding the ceiling $\bar{h}$ results in negative innovation. A better form is common:

$$S(h) = \frac{1}{1 + \frac{h}{\bar{h}}} \tag{2.81}$$

In this form, $\bar{h}$ is called the "half-saturation constant," as $S = .5$ when $h = \bar{h}$. In this form, S is positive for all positive values of h, decreasing monotonically.

2.6 RECAP

This chapter has developed many basic bioeconomic interactions among fish and fishers – more generally, prey and predator or resource and people. The basic bioeconomic model can be recognized as the "Gordon-Scott fishery" [32, 80]. It is used in many expositions, including resource economics generally – for example, Conrad [12], and in fisheries treatments specifically: Clark [9], van den Bergh [90], Grafton et al [34], and Mangel [59]. These works and many other typically go much further into the description of biological populations, resonating well with the subsequent Chapters 3 and 4 herein.

Critical issues reveal the necessity of sustaining the ecosystem; the biological populations hosted; and the economic interactions with people. Achieving this set of outcomes will require a high degree of practicality in diverse cases. A realistic view of local/private incentives, ownership, and the common object is needed, as is a careful assessment of individual objectives and motivations. The simple model used here illumines the multiplicity of criteria that might be relevant: the biomass, the harvest, the rent, the jobs. All have found their their place in practical systems.

Several works address operational issues encountered in fisheries management. Included are Walters and Martell [91], Hillborn and Walters [39], and Clark [10]. As used here, fisheries represent a case study of the more general class of living renewables.

The critical issue of management of common-pool resources is fundamental to the considerations introduced here. The reader is referred to the important contributions of Ostrom et al. [71, 70] and Sandler [79], for a blend of theory and experiment. These are important contributions to natural resource management, touching fundamental social science issues broadly.

2.7 PROGRAMS

The following programs illustrate the ideas in these lectures:

Fish1.4.xls (fish1_4.m) simulates the dynamics of B and E under open-access conditions.

Fish1.4-RAND.xls adds independent stochastic disturbances to G and dE/dt. Both time series are autocorrelated.

Fish1.4-Cubic.xls adds depensation to G and a minimum viable biomass parameter $\underline{B}$.

Fish1.5-RAND.xls adds dynamics of technological innovation to **Fish1.4-RAND.xls**.

Fish1.6.xls (fish1_6.m) computes the equilibrium for the costly harvesting of logistic growth, with intertemporal considerations of interest rate, as in Section 2.4.2.

2.8 PROBLEMS

1 Logistic growth function:

(a) Express the logistic growth function, Equation 2.2, in terms of the departure from carrying capacity $\epsilon \equiv K - B$.

(b) Show that for small ϵ,

$$\frac{d\epsilon}{dt} = -g\epsilon \tag{2.82}$$

2 Logistic growth, no harvesting. Find the waiting time T_p until peak growth rate occurs for the following:

(a) $g = .08/\text{yr}$, $B_0/K = 5\%$

(b) An endangered species with slow growth, $g = .04/\text{yr}$, and low abundance $B_0/K = 10\%$

(c) An invasive species at low initial abundance $B_0/K = 7\%$, and a doubling time of 20 years

3 For the steady-state fishery:

$$G = gB(1 - B/K) \tag{2.83}$$

$$H = hEB \tag{2.84}$$

$$\pi = pH - cE \tag{2.85}$$

Plot harvest as a function of price p, for the open-access (Rent $= 0$) and for the closed-access (Rent-maximizing) cases. Use these parameters: $(K, c, g, h) = 1$. Explain your results.

4 Consider a fishery with growth given as $G = g(1 - B/K)$ and all other relations unchanged from the case in Problem 3. (Be careful about G here; it is not hard, just different.)

(a) Solve for B, H, E, and rent under open access.

(b) Repeat for the closed-access case.

(c) Plot H versus p as in Problem 3.

(d) Compare the two management regimes.

5 Simulation exercises with the **FISH1.4** model. This model solves the simple logistic fishery

$$G = gB(1 - B/K) \tag{2.86}$$

$$H = hEB \tag{2.87}$$

$$\pi = pH - cE \tag{2.88}$$

with dynamics

$$\frac{dB}{dt} = G - H \tag{2.89}$$

$$\frac{dE}{dt} = \nu\pi \tag{2.90}$$

The standard parameter setup is

Time	dt	1
Growth	g	2
	K	1
Harvesting	h	1
	ν	0.5
Value	p	5
	c	3

Confirm: With all other parameters fixed, what values of ν lead to (a) orderly (monotonic) approach to equilibrium; (b) oscillatory approach to equilibrium; (c) eternal ocsillations; and (d) instability or extinction.

6 Repeat the simulations from Problem 5, with a stochastic disturbance ϵ affecting the growth function:

$$G = gB(1 - B/K)(1 + \epsilon) \tag{2.91}$$

with ϵ a random number with zero mean, variance σ^2. Use the three values of ν found in Problem 5 and values of autocorrelation $\rho = 0, .50, .75, .90$, and $.95$. Study and report the effect of disturbance size σ in each of the cases.

7 Add depensation to Problem 5 by altering the growth function to be

$$G = g(1 - B/K)(B - 0.25) \tag{2.92}$$

Restudy Problem 5. Are your conclusions about parameter ranges, extinction possibility, etc. altered?

8 Repeat the simulations from Problem 7, with a stochastic disturbance ϵ affecting the growth function:

$$G = g(1 - B/K)(B - 0.25)(1 + \epsilon) \tag{2.93}$$

with ϵ a random number with zero mean, variance σ^2. Use the three values of v found above (Problem 7) and values of autocorrelation $\rho = 0, .50, .75, .90,$ and $.95$. Study and report the effect of disturbance size σ in each case.

9 For the standard fishery:

$$G = gB(1 - B/K) \tag{2.94}$$

$$H = hEB \tag{2.95}$$

$$\pi = pH - cE \tag{2.96}$$

with parameters $g = 1, h = 1, K = 1, p = 5, c = 3$. Currently, this fishery is open access and at steady state. There are numerous small fishing businesses. It is decided to change to a controlled-access regime and to maximize rent. The essence of the problem is reducing fishing effort – that is, all fishers will not continue to fish at the same rate.

  (a) One proposal is to *sell permits* that allow access to the fishery. One permit would allow .01 unit of fishing effort per year. How many permits need to be sold each year, and what is their price?
  (b) Instead, *fish will be taxed* at the dock, with fishers paying a tax, t, per unit of harvest. What value of tax is needed?
  (c) Instead, the *government will do the fishing* and issue no permits. Current fishers will be offered government jobs, either fishing or teaching ecology courses in the local high school. Both will be paid the same wage, that of a current fisher. What proportion of the current fishers will become teachers?

10 Consider the following fishery:

$$G = gB^2(1 - B/K) \tag{2.97}$$

$$H = hEB \tag{2.98}$$

$$\pi = pH - cE \tag{2.99}$$

This is the standard model except that the growth function G is different. (Be careful about this before starting!) Parameters are $g = 2, h = 1, K = 1, p = 5, c = 3$.

  (a) Sketch the growth function G versus B. Pay attention to the slope and curvature near the origin.
  (b) Is there depensation?
  (c) What is the maximum sustainable yield? And what is the associated value of B?
  (d) Find $B, H,$ and E under controlled-access conditions.

11 Given the usual harvesting and rent relations

$$H = hEB \tag{2.100}$$

$$\pi = pH - cE \tag{2.101}$$

Consider the following growth functions:

(i) $G = \frac{1}{B}$

(ii) $G = \frac{1}{B} - \frac{1}{K}$

(iii) $G = 1 - e^{-B}$

(iv) $G = \sqrt{B}$

(v) $G = Be^{-B}$

(vi) $G = g\,\sin\left(\frac{B}{K}\Pi\right)$ (Here Π is the geometry constant 3.1416, as distinct from π, which indicates rent.)

(vii) $G = gB/\left(1 + \frac{B}{K}\right)$

For each case, find B, E, H, π for

(a) the open-access equilibrium.

(b) the closed-access equilibrium.

(c) the closed-access intertemporal equilibrium for $r = 0$ and $c = 0$.

(d) the closed-access intertemporal equilibrium for $r > 0$ and $c = 0$.

(e) the closed-access intertemporal equilibrium for $r > 0$ and $c \neq 0$.

Express your answers in terms of the parameters p, c, g, h, K, r, and, where useful, $B_0 \equiv c/ph$.

12 A fishery is characterized by

$$G = gBe^{-B}$$

$$H = hEB$$

Parameters: $(g, h, p, c) = (2, 3, 5, 2)$.

(a) Find the open-access steady state: E, B, H, π.

(b) Find the closed-access steady state: E, B, H, π.

(c) Find the MSY steady state: E, B, H, π.

13 For the standard fishery,

$$G = gB\left(1 - \frac{B}{K}\right)$$

$$H = hEB$$

Parameters: $(g, h, p, c, K) = (1, 1, 4, 1, 1)$. This fishery is now operating in open-access steady state. It is a concern that the level of B is too low, risking extinction. Therefore, it is desired to (i) close the access and keep the *effort constant*; and simultaneously, (ii) *restrict the harvesting technology*.

(a) What is the open-access status quo, (E, B, H, π)?

(b) What value of h is needed to move this fishery to the desired new steady state at MSY? (Note that there is no change in effort.)

(c) After making this change in h, what are the new steady-state values of (E, B, H, π)? Which have changed and in what direction?

(d) The rent will be distributed equally among the fishers. What is the payment per unit of effort?

(e) What is the *effective* compensation per unit of effort?

(f) Comment: The harvest is larger, the technology is cruder, the effort is the same, but the fishers are getting more money. Is this right, and why?

14 Given

$$H = hEB$$

$$\pi = pH - cE$$

and *any* growth function $G = G(B)$. Show that in steady state the competitive (open-access) equilibrium biomass is independent of the form of G:

$$B = \frac{c}{ph}$$

15 A fishery is described by the growth function

$$G(B) = g\left[1 - e^{-\alpha B}\right] \tag{2.102}$$

with constants g and α. Under costless production, what value of interest rate r will result in extinction?

16 A certain marine mammal population has a doubling time of 20 years in the absence of harvesting. The prevailing interest rate is 5% per year. Assuming costless production, is this species endangered?

17 A fishery has

$$G = g\frac{B}{(1 + B/K)} \tag{2.103}$$

$$H = hEB \tag{2.104}$$

with $g = .8$ and $K = 2$. If effort is held fixed at $E = 5$, what value of technology h will result in extinction?

18 Here is a fishery to be operated in steady state:

$$G = g(1 - B/K)B^2 \tag{2.105}$$

$$H = hEB \tag{2.106}$$

$$\pi = pH - cE \tag{2.107}$$

(Notice that the last term in G is squared; this is depensatory.) Parameters: $(g, h, K, p, c) = (2, 1, 1, 5, 3)$.

(a) In open-access mode, steady state, what are (B, E, H, π)?

(b) Keeping E fixed from (a): What is the maximum value of technology h beyond which biological extinction is guaranteed? (Hint: Consider H and G; this is a biological question, not an economic one.)

(c) Ignore (b): We wish to add a tax to the price of fish sold in order to achieve $B = 0.9$. (This money will be used to support public education.) What value of tax is needed, and how much total tax is to be collected?

(d) Ignore (b) and (c): The cost of effort will rise due to a new minimum wage law under study: c becomes $c + d$. What is the relation between jobs E and the extra wage d?

(e) What is the maximum value of d above which we will have economic extinction – that is, unprofitable fishing for all $0 < B < K$?

19 Explorers encountered a new fishery at high abundance a few years ago. After a brief period of costless mining, they are now operating it sustainably at $B = .35K$. Parameters: $p = 5$, $c = 0$, $r = .05$. It is thought that the growth function is

$$G = g\frac{B}{(1 + B/K)} \tag{2.108}$$

(Notice that B can be higher than K here.)

(a) What is the value of g?

(b) What change in interest rate would lead to extinction?

(c) A worldwide depression causes r to shrink to .03. What will happen to B? What change in sustainable harvest will occur? (Continue to assume that $c = 0$.)

20 Same as Problem 19, except that there is new information about the growth function G. It is now thought that

$$G = g\left(1 - \frac{B}{K}\right)B^2 \tag{2.109}$$

(Notice that the last term is squared.)

(a) What is the value of g?

(b) Is this fishery operating in a depensatory regime?

21 A fishery is described by

$$G = B(1 - B) \tag{2.110}$$

$$H = hB\sqrt{E} \tag{2.111}$$

$$\pi = 2H - 3E \tag{2.112}$$

(Notice the $\sqrt{E}$ term.) The parameter h represents harvesting technology.

(a) Find the MSY point and the values of H, E, B, π there.

(b) Find the the values of H, E, B, π for an open-access fishery.

(c) Find the maximum sustainable rent possible and the values of H, E, B there.

(d) Plot steady-state rent versus h for both open and closed-access cases.

22 (See Section 2.4.2.) A fishery is described by

$$G = gB \left(1 - \frac{B}{K} \right)$$ (2.113)

$$H = hEB$$ (2.114)

$$\pi = pH - cE$$ (2.115)

A pioneer has encountered this fishery at carrying capacity. She proposes a one-time large harvest, reducing the biomass to $B1$; investing the proceeds forever at interest rate r; and thereafter, harvesting sustainably under controlled-access conditions. (Note: The cost of harvesting is not zero.) Parameters: $[K, g, h] = [1, .06, 1]$; $[p, c, r] = [5, 3, .10]$.

(a) Compute $B1$.

(b) This fishery has been operating sustainably at $B1$ for several years. Suddenly there is a financial crisis, and r drops permanently to zero. Predict the effect on the fishery under controlled access. Use mathematical reasoning, Also explain the result graphically and in common sense terms.

(c) Ignore (b). Is it possible that technological advances will justify extinction in the controlled-access case? If yes, under what conditions? Are they met here?

23 Consider the fishery as in Problem 4, with $H = hEB$ and $G = g(1 - B/K)$, unchanged; *but* with a minimum viable biomass $\underline{B}$ such that $G = 0$ for $B < \underline{B}$. (In other words, G crashes to zero if B falls below $\underline{B}$.) Assume the rent-maximixing (closed-access) solution from Problem 4 and that this fishery is in steady state.

(a) Sketch the growth function $G(B)$. In what range of B is there depensation?

(b) There is a food shortage; fish prices rise. What value of price will cause the extinction of this fishery?

(c) There is no food shortage, but suddenly there is a stock market boom, and it is suggested that the fishery regulation board harvest some extra fish, sell them, and buy industrial stocks with the cash. The expected annual return on this investment is r. Compute the new optimal level $B1$. (Hint: First express steady-state rent π_s in terms of B; then use Equation 2.48 from the text.)

(d) At what value of r will the fishery become extinct?

3 Stage-Structured Populations

Here we continue with the exposition of living resources in Quadrants 3 and 4. In the previous chapter, we introduced a single biomass scalar B and described its bioeconomic interactions. The natural extension of this is to distinguish different *life stages* both biologically and economically. The biological description will be in terms of *numbers of individuals per stage*. The size, weight, etc. of each stage is treated as a biological fact.

We will carry forward the same approach to harvesting, but consider the realistic possibility of directing effort toward certain life stages only. In addition to the life stage realism, this reveals the important opportunity to describe *recruitment* – that is, the number of individuals born into the population. There are two limiting extremes: (1) where recruitment depends on the number of mature individuals; and (2) where recruitment is effectively controlled by other things. Needless to say, there is a continuum, and one needs to know which regime is controlling and to manage accordingly.

The case of fish populations and management continues to be used to illustrate ideas.

3.1 POPULATION STRUCTURE

In the previous chapter, we examined the simple model with a single biological state – the scalar biomass B. Here we expand the description to include natural systems with greater biological complexity stemming from growth and development of specific organisms through different natural *life stages*. Hence, stage-structured descriptions. As we shall see, it will be convenient to generalize the scalar biomass to a *biomass vector* $\{B\}$, containing the inventories of all the (scalar) biomass by life stage.

3.1.1 A Two-Stage System

We begin with a simple example, with two stages:

Y: number of young fish
A: number of adult fish

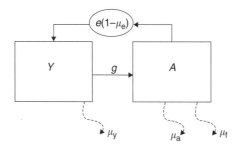

Figure 3.1. Box model of the two-stage population. Equations 3.1 and 3.2 govern (in matrix form, Equation 3.3). Growth (*g*), reproduction (*e*), mortality (μ), and harvest (fishing mortality μ_f) are all first-order processes.

Adopting a discrete-time description, we can account for several vital rates:

g: growth (aging, development) rate
μ: mortality rate
μ_f: fishing mortality (harvesting) rate
e: reproduction rate

These are illustrated in Figure 3.1. The equations implied are

$$Y^{k+1} = Y^{k+1} - gY^k - \mu_y Y^k + e(1 - \mu_e)A^k \tag{3.1}$$

$$A^{k+1} = A^{k+1} + gY^k - \mu_a A^k - \mu_f A^k \tag{3.2}$$

where the superscript indicates time level. We imagine a discretization that is natural – for example, annual or seasonal – for the population being described and, ultimately, for the harvesters or managers. In this example, fishing mortality is directed at the adult, but not the young, segments of the population. Similarly, reproduction is relevant for the adults only. The reproduction eA^k is a source for Y, with net survival $(1 - \mu_e)$; it occurs without a time lag. The harvest $H \equiv \mu_f A^k$ is a sink for A. The growth term gY^k is a transfer between them.

In matrix form, we have

$$\left\{ \begin{array}{c} Y \\ A \end{array} \right\}^{k+1} = \left[\begin{array}{cc} (1 - g - \mu_y) & e(1 - \mu_e) \\ g & (1 - \mu_a - \mu_f) \end{array} \right] \left\{ \begin{array}{c} Y \\ A \end{array} \right\}^{k} \tag{3.3}$$

The matrix in this equation is the *Leslie matrix* [*L*]. In this case, it accounts for the entire dynamic of this simple population.

Equation 3.3 represents a homogeneous system. It has no exogenous forcing; all the dynamic is internal to it. Solutions may be sought that have the self-replicating property

$$\left[L \right] \left\{ \begin{array}{c} Y \\ A \end{array} \right\} = \lambda \left\{ \begin{array}{c} Y \\ A \end{array} \right\} \tag{3.4}$$

or equivalently,

$$\left[L - \lambda I \right] \left\{ \begin{array}{c} Y \\ A \end{array} \right\} = 0 \tag{3.5}$$

Substituting into Equation 3.3, we get

$$\left\{ \begin{array}{c} Y \\ A \end{array} \right\}^{k+1} = \lambda \left\{ \begin{array}{c} Y \\ A \end{array} \right\}^{k} \tag{3.6}$$

Here λ is a scalar growth rate. As hypothesized, it transforms today's population structure into the same structure tomorrow, magnified but otherwise unchanged.

The scalar λ and the vector $\{Y, A\}$ constitute a coupled pair. For nonsingular Leslie matrices of order $N \times N$, we expect to find exactly N such pairs. They have the property of the singularity of $[L - \lambda I]$:

$$\mid L - \lambda I \mid = 0 \tag{3.7}$$

This condition defines λ. Some points:

- λ's that satisfy this singularity condition represent the growth rate of self-replicating modes

$$\left\{ \begin{array}{c} Y \\ A \end{array} \right\}^{k+1} = \lambda \left\{ \begin{array}{c} Y \\ A \end{array} \right\}^{k} \tag{3.8}$$

- Each mode will have its own population structure, in this case the ratio of young to adult fish, Y/A.
- It is conventional to normalize these vector modes – for example, making $A = 1$.
- For an $N \times N$ Leslie matrix, under ideal conditions we can expect at most N separate λ's and corresponding population modes.
- The largest λ represents the fastest growth rate in the system; the corresponding population mode will dominate at large time, as it will outpace all the rest.
- The mode associated with the largest λ is the *Stable Population Mode*. Normalized populations would exhibit this mode at large time.
- The modes will be independent of each other; the population at any time can be expressed in terms of these.

In this example, we have

$$[L - \lambda I] = \left[\begin{array}{cc} (1 - g - \mu_y - \lambda) & e(1 - \mu_e) \\ g & (1 - \mu_a - \mu_f - \lambda) \end{array} \right] \tag{3.9}$$

and its determinant is readily expanded to

$$|L - \lambda I| = (1 - g - \mu_y - \lambda)(1 - \mu_a - \mu_f - \lambda) - e(1 - \mu_e)g = 0 \tag{3.10}$$

This is a quadratic equation with two roots λ. The modes are obtained from the Leslie equation itself:

$$(1 - g - \mu_y - \lambda)Y + e(1 - \mu_e)A = 0 \tag{3.11}$$

$$gY + (1 - \mu_a - \mu_f - \lambda)A = 0 \tag{3.12}$$

These provide two different expressions; but for the special values of λ that satisfy Equation 3.7, the system is singular and they are equivalent.

$$\frac{Y}{A} = -\frac{e(1 - \mu_e)}{(1 - g - \mu_y - \lambda)} \tag{3.13}$$

$$= -\frac{(1 - \mu_a - \mu_f - \lambda)}{g} \tag{3.14}$$

For these parameters:

$g = .4$
$e(1 - \mu_e) = 10.$
$\mu_y = .2$
$\mu_a = .2$
$\mu_f = 0.$

we have

$$\lambda_1 = 2.61 \quad \frac{Y}{A} = 4.53 \tag{3.15}$$

$$\lambda_2 = -1.41 \quad \frac{Y}{A} = -5.53 \tag{3.16}$$

So, the stable population mode has 4.53 more young than adult fish; and it increases by the factor 2.61 each year. The second mode also grows without bound, but ultimately it cannot keep up. Its signature at early time would be an oscillation in the population structure, as λ and Y/A are negative for this mode.

In Figure 3.2, we illustrate this case. Following a jittery start, the population structure settles in on the stable population mode quickly. The annual increment in adult biomass would be $\Delta \ln A = \ln \lambda_1$, roughly 0.96 in this case.

Continuing the example, we can add fishing mortality $\mu_f = 0.5$. Going back to the quadratic equation 3.10, we get

$$\lambda_1 = 2.35 \quad \frac{Y}{A} = 5.13 \tag{3.17}$$

$$\lambda_2 = -1.65 \quad \frac{Y}{A} = -4.88 \tag{3.18}$$

The fishing mortality has decreased both the growth rate and the relative abundance of adult fish in the dominant mode.

Programs **FishYA.xls** and **ex2box.m** illustrate these points. The stable population mode and its growth rate emerge early and dominate the dynamic in simple simulations of this unstably growing population.

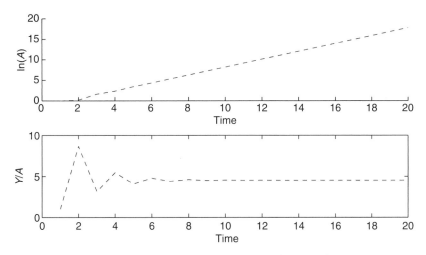

Figure 3.2. The stable population mode for the two-box model; $\mu_f = 0$.

3.1.2 Biomass Vector

Here and below, it is convenient to identify the *biomass vector* {B} as containing all the life stages – in this case,

$$\left\{ B \right\} \equiv \left\{ \begin{matrix} Y \\ A \end{matrix} \right\} \tag{3.19}$$

In these terms, Equation 3.3 is generalized to

$$\left\{ B \right\}^{k+1} = \left[L \right] \left\{ B \right\}^{k} \tag{3.20}$$

and we have homogeneous solutions in the form

$$\left\{ B \right\}^{k+1} = \lambda \left\{ B \right\}^{k} \tag{3.21}$$

$$\left[L \right] \left\{ B \right\} = \lambda \left\{ B \right\} \tag{3.22}$$

3.2 RECRUITMENT

Recruitment R is the rate of addtion to the youngest stage. In the previous example, it is linear in A:

$$R = e(1 - \mu_e)A \tag{3.23}$$

The number of recruits per adult fish, R/A, is constant; there is no density dependence in this ratio. In the linear model described above, the population has two choices, depending on the vital rates: grow without bound, or collapse to zero. There is nothing against which to balance a carrying capacity or a finite steady state.

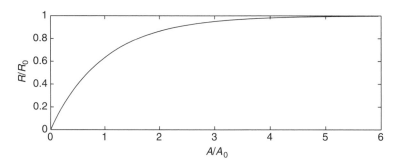

Figure 3.3. Candidate recruitment rate function

At high A, recruitment may saturate for any number of practical reasons: habitat, predation, food supply, etc. A candidate representation would be

$$R(A) = R_0(1 - e^{-A/A_0}) \tag{3.24}$$

which is drawn in Figure 3.3. At low abundance, R is linear in A:

$$R \simeq R_0 \frac{A}{A_0} \tag{3.25}$$

and at high abundance, R saturates at its upper limit:

$$R \simeq R_0 \tag{3.26}$$

Effectively, these processes, external to the population, set a carrying capacity for it by limiting recruitment.

With this modification, it is convenient to restate the Leslie relations with recruitment exogenous

$$\left\{ \begin{array}{c} Y \\ A \end{array} \right\}^{k+1} = \left[\begin{array}{cc} (1 - g - \mu_y) & 0 \\ g & (1 - \mu_a - \mu_f) \end{array} \right] \left\{ \begin{array}{c} Y \\ A \end{array} \right\}^{k} + \left\{ \begin{array}{c} R \\ 0 \end{array} \right\} \tag{3.27}$$

These are illustrated in Figure 3.4. They have a steady-state solution:

$$Y = \left[\frac{1}{\mu_y + g} \right] R$$

$$A = \left[\frac{g}{\mu_a + \mu_f} \right] Y = \left[\frac{g}{\mu_a + \mu_f} \right] \left[\frac{1}{\mu_y + g} \right] R \tag{3.28}$$

These solutions are steady and proportional to R. The case $\mu_f = 0$ is the no-harvest case; it sets the upper limit on $A = \bar{A}$:

$$\bar{A} = \left[\frac{g}{\mu_a} \right] \left[\frac{1}{\mu_y + g} \right] R \tag{3.29}$$

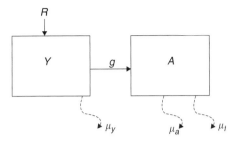

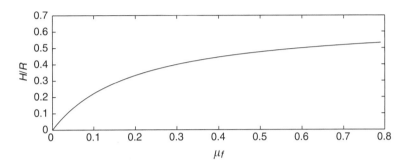

Figure 3.4. Box model of the two-stage population. Equation 3.27 governs. Recruitment R is displayed as exogenous.

Figure 3.5. Steady-state harvest per recruit versus fishing mortality. Parameters: $g = 0.4$; $\mu_y = 0.2$; $\mu_a = 0.2$. For these parameters, $\overline{H}/R = .533$.

Similarly, the lower limit $A = \underline{A}$ is set at the maximum of μ_f: $\mu_f + \mu_a = 1$:

$$\underline{A} = \left[\frac{g}{1}\right]\left[\frac{1}{\mu_y + g}\right] R \tag{3.30}$$

The harvest is $\mu_f A$:

$$H = \left[\frac{\mu_f g}{\mu_a + \mu_f}\right]\left[\frac{1}{\mu_y + g}\right] R \tag{3.31}$$

and the harvest is maximum at $\underline{A}$, with $\mu_f + \mu_a = 1$:

$$\overline{H} = \left[\frac{(1 - \mu_a)g}{\mu_y + g}\right] R \tag{3.32}$$

In Figure 3.5, the harvest is plotted versus μ_f.

3.2.1 Rent

With the above, we can recover the sense of rent in this fishery. The effort here, μ_f, is the product of harvesting technology h and effort E in the previous chapter:

$$H = \mu_f A = hEA \tag{3.33}$$

and hence with the cost of the fishing mortality c, we have rent:

$$\pi = pH - c\mu_f \tag{3.34}$$

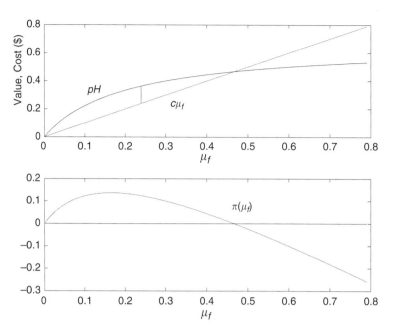

Figure 3.6. Steady-state rent versus fishing mortality. Income pH and expenses $c\mu_f$ are plotted. Their difference is π. Parameters: $g = 0.4$; $\mu_y = 0.2$; $\mu_a = 0.2$; $pR = 1$; $c = 1$.

Putting these together, we have

$$\pi = p \left[\frac{\mu_f g}{\mu_a + \mu_f} \right] \left[\frac{1}{\mu_y + g} \right] R - c\mu_f \tag{3.35}$$

These are plotted in Figure 3.6. The gap between cost and value is the rent.

As in the simpler case discussed in Chapter 2, where we had only a single biomass variable B: Here also there are many possible steady states, each depending on the choice of μ_f. Two standard cases are worth note. The first is the open-access case, wherein μ_f is allowed to rise in response to positive π. The stable equilibrium would be at $\pi = 0$ and high μ_f:

$$\mu_f + \mu_a = \frac{pg}{c} \frac{R}{(\mu_y + g)} \tag{3.36}$$

The equilibrium μ_f responds positively to increases in R or p or to decreases in c. Fishing mortality cannot exceed $\mu_f = 1 - \mu_a$; so for very high R or p, this zero-rent condition cannot be reached. That happens when

$$\frac{pg}{c} \frac{R}{(\mu_y + g)} > 1 \tag{3.37}$$

in which case positive rent would still accrue in open access and μ_f would be at the maximum compatible with steady state.

Alternatively, we can find the point of maximum rent, achievable if access is closed and μ_f can be controlled. Setting $d\pi/d\mu_f = 0$ and solving for μ_f gives the maximum

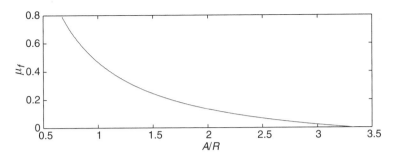

Figure 3.7. A versus μ_f from Equations 3.28 or 3.39. This translates Figure 3.6 into Figure 3.8.

rent condition

$$\left(\mu_f + \mu_a\right)^2 = \mu_a \frac{pg}{c} \frac{R}{(\mu_y + g)} \tag{3.38}$$

By comparison, we can see that the closed-access case is achieved at lower μ_f than in the open-access case.

Rent versus A

An equivalent set of plots and relations are useful in terms of *adult biomass*. The steady-state relation between A and μ_f is given in Equation 3.28, rearranged:

$$\mu_f + \mu_a = \frac{gR}{A(\mu_y + g)} \tag{3.39}$$

which is plotted in Figure 3.7. The Harvest $H = \mu_f A$:

$$H = \frac{gR}{\mu_y + g} - \mu_a A \tag{3.40}$$

that is, equilibrium H is linear in A. The rent $\pi = pH - c\mu_f$ in terms of A is

$$\pi(A) = \left[p - \frac{c}{A}\right]\left[\frac{gR}{\mu_y + g} - \mu_a A\right] \tag{3.41}$$

Rent versus A is plotted in Figure 3.8. $\pi(\mu_f)$ and $\pi(A)$ represent the same steady relations. Comparison of $\pi(A)$ here, with the simple logistic result $\pi(B)$ in Chapter 2 (Equation 2.15, Figure 2.5), is invited.

3.2.2 A Different View: Steady Harvest

It is instructive to treat harvest $H \equiv \mu_f A$ directly and reformulate Equations 3.27 thus:

$$\begin{Bmatrix} Y \\ A \end{Bmatrix}^{k+1} = \begin{bmatrix} (1 - g - \mu_y) & 0 \\ g & (1 - \mu_a) \end{bmatrix} \begin{Bmatrix} Y \\ A \end{Bmatrix}^{k} + \begin{Bmatrix} R \\ -H \end{Bmatrix} \tag{3.42}$$

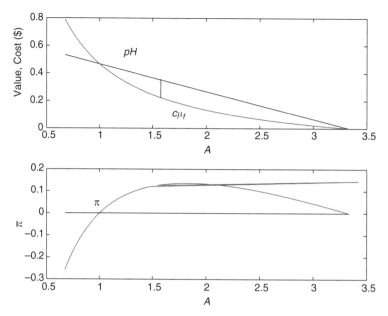

Figure 3.8. Steady-state rent versus adult biomass A; same as Figure 3.6, but the horizontal axis is A. Income pH, expenses $c\mu_f$, and rent π are plotted. (Compare Figure 2.5, where $\pi(B)$ is plotted for the simpler logistic model.)

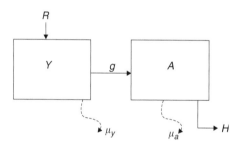

Figure 3.9. Box model of the two-stage population. Equation 3.42 governs. Recruitment R and harvest H are both displayed as exogenous.

This is illustrated in Figure 3.9.

We have an algebraic system in which both R and H appear as primary quantities and the effortlike variable μ_f is implied but not stated. In these terms, the steady-state solution is

$$Y = \left[\frac{1}{\mu_y + g}\right] R$$

$$\frac{H}{R} = \left[\frac{g}{\mu_y + g}\right] - \mu_a \frac{A}{R} = \mu_a \left[\frac{\overline{A}}{R} - \frac{A}{R}\right]$$

$$\mu_f = \left[\frac{g}{\frac{A}{R}(\mu_y + g)}\right] - \mu_a \tag{3.43}$$

(The latter is simply $\mu_f = H/A$.) These relate the economic variables to A; they are equivalent to those relating to μ_f (Equation 3.28), with a change to H as a

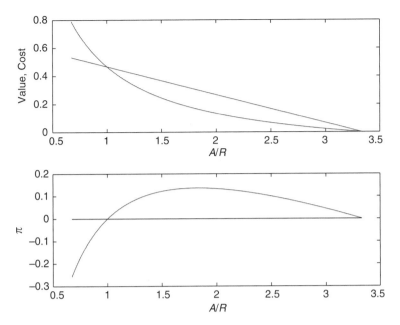

Figure 3.10. Duplication of Figure 3.8, via Equation 3.43.

primary variable. Their plot is the same as presented in Figure 3.8. It is duplicated in Figure 3.10.

3.3 DYNAMICS: EXOGENOUS R

Following the previous section with R set exogenously, we have the Leslie matrix as in Equation 3.27. Its solution will have two parts: a steady-state particular solution, already explored, and a dynamic homogeneous solution in terms of eigenmodes. The latter will rely on the eigenstructre of the Leslie matrix with recruitment removed:

$$\begin{vmatrix} (1 - g - \mu_y - \lambda) & 0 \\ g & (1 - \mu_a - \mu_f - \lambda) \end{vmatrix} = 0 \qquad (3.44)$$

The solution is especially simple here:

$$(1 - g - \mu_y - \lambda)(1 - \mu_a - \mu_f - \lambda) = 0 \qquad (3.45)$$

which gives

$$\begin{aligned} \lambda_1 = 1 - g - \mu_y \qquad & Y/A = \frac{\mu_a + \mu_f - \mu_y - g}{g} \\ \lambda_2 = 1 - \mu_a - \mu_f \qquad & Y/A = 0 \end{aligned} \qquad (3.46)$$

Both modes represent decay (the instability due to endogenous reproduction is removed). For the parameters $(g, \mu_y, \mu_a, \mu_f) = (.4, .2, .2, 0)$, we get

$$\lambda_1 = 0.4 \qquad Y/A = -1$$
$$\lambda_2 = 0.8 \qquad Y/A = 0 \qquad\qquad (3.47)$$

The first mode decays the fastest toward steady state; it oscillates between Y and A. The second mode decays more slowly and would persist the longest following a departure from steady state. It is a pure A mode; Y is zero. These results are for the undisturbed system, $\mu_f = 0$. If we add fishing mortality at $\mu_f = .3$, for example, we get

$$\lambda_1 = 0.4 \qquad Y/A = -.25$$
$$\lambda_2 = 0.5 \qquad Y/A = 0 \qquad\qquad (3.48)$$

The details of the temporal response are altered, in this case toward faster adjustment toward steady state. Of course, the steady state itself is different, due to the change in μ_f, as described above (Equation 3.28; Figures 3.5 and 3.7).

The Excel program **FishYA_R** simulates this system. The student is asked to verify (a) the unstable growth of Leslie modes at low abundance; and (b) the stable decay to a steady state at high abundance. The candidate recruitment function in that program is

$$R(A) = R_0 \left[1 - e^{-A/A_0} \right] \qquad\qquad (3.49)$$

3.4   THE FISH FARM

Here we retain the stage-structured analysis, but abandon the context of wild population structure and the attempt to control fishing mortality. Instead, it is assumed here that the harvest itself can be controlled and that it can be directed at specific stages with precision. The dynamic is as in Equation 3.3, with exogenous harvest H_Y and H_A directed at the young and adult stages, respectively:

$$\begin{Bmatrix} Y \\ A \end{Bmatrix}^{k+1} = \begin{bmatrix} (1 - g - \mu_y) & e(1 - \mu_e) \\ g & (1 - \mu_a) \end{bmatrix} \begin{Bmatrix} Y \\ A \end{Bmatrix}^{k} - \begin{Bmatrix} H_Y \\ H_A \end{Bmatrix}^{k} \qquad (3.50)$$

Notice that we have endogenous recruitment here. Unless there is something to constrain the population, it will grow without bound. A carrying capacity will need to be imposed somehow. We will impose a space limitation: at least σ_y space per young fish, σ_a per adult:

$$S = \sigma_y Y + \sigma_a A \leq \overline{S} \qquad\qquad (3.51)$$

which must govern at all times. (Analogous forms would occur for limitations on other resources: light, food, etc.) It is also understood that harvest and biomass are nonnegative at all times:

$$(H_y, H_a, Y, A) \geq 0 \tag{3.52}$$

There are no exogenous inputs here. Essentially, this system is not "seeded" with Y or A from any other source. Its reproduction is entirely endogenous.

In steady state, we have

$$\begin{bmatrix} (-g - \mu_y) & e(1 - \mu_e) \\ \\ g & (-\mu_a) \end{bmatrix} \begin{Bmatrix} Y \\ A \end{Bmatrix} = \begin{Bmatrix} H_y \\ H_a \end{Bmatrix} \tag{3.53}$$

Solving these for the harvest and collecting the other relations, we have five inequalities:

$$
\begin{aligned}
H_y &= -(g + \mu_y)Y + e(1 - \mu_e)A \geq 0 \\
H_a &= \quad\quad gY - \mu_a A \quad\quad \geq 0 \\
S &= \quad\quad \sigma_y Y + \sigma_a A \quad\quad \leq \overline{S} \\
&\quad\quad\quad\quad\quad Y \quad\quad\quad\quad \geq 0 \\
&\quad\quad\quad\quad\quad A \quad\quad\quad\quad \geq 0
\end{aligned}
\tag{3.54}
$$

This set of inequalities is linear in the decisions Y, A. Together, they bound a set of feasible solutions, illustrated in Figure 3.11. Any of the Y, A points within the feasible set are possible. For example, an economic actor is likely to define value in terms of the price gotten from sale of the harvest, p_y, p_a, and the costs associated with maintaining the fish farm population c_y, c_a:

$$\pi = p_y H_y + p_a H_a - c_y Y - c_a A \tag{3.55}$$

It is easy to eliminate H_y, H_a in favor of Y, A:

$$\pi = p_y \left[-(g + \mu_y)Y + e(1 - \mu_e)A \right] + p_a \left[gY - \mu_A A \right] - c_y Y - c_a A \tag{3.56}$$

After some trivial rearrangement, we get

$$\pi = \left[-p_y(g + \mu_y) + p_a g - c_y \right] Y + \left[p_y e(1 - \mu_e) - p_a \mu_A - c_a \right] A \tag{3.57}$$

Such an economic actor would maximize π, subject to the constraints (Equation 3.54). This would isolate one operating point within the feasible space, defining population

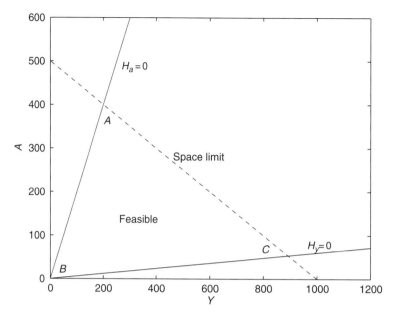

Figure 3.11. Fish farm: feasible operating options with endogenous linear recruitment. The constraints $H_y \geq 0$ and $H_a \geq 0$ are shown, plus the carrying capacity represented by the space constraint. Together, the three constraints confine the feasible region, with extreme points A, B, and C. Parameters are as in the example.

(Y, A), harvest (H_y, H_a), space required S, and profit π. This objective is linear in (Y, A), as are all the constraints. We therefore know that the optimal solution will lie on one of the extreme corners, A, B, or C in Figure 3.11, except in degenerate cases.

3.4.1 Example

Let $(\mu_Y, \mu_A, e(1 - \mu_e), g) = (.2, .2, 10, .4)$ and $(\sigma_y, \sigma_a, \overline{S}) = (1, 2, 1000)$. Then we have

$$H_y = -.6Y + 10A \geq 0$$
$$H_a = .4Y - .2A \geq 0$$
$$S = Y + 2A \leq 1,000 \tag{3.58}$$
$$Y \geq 0$$
$$A \geq 0$$

Adding the financial parameters $(p_y, p_a, c_y, c_a) = (3, 5, 4, 3)$, we get

$$\pi = -3.8Y + 26A \tag{3.59}$$

π is to be maximized subject to the five constraints (Equation **??**).

This maximum is found at point A, where $H_a = 0$ and $S = \bar{S}$. The constraints on nonnegative H_a and on limited S *bind*; the other constraints are *slack*. The complete solution is

$$Y = 200$$
$$A = 400$$
$$H_y = 3{,}880 \qquad (3.60)$$
$$H_a = 0$$
$$S = 1{,}000$$
$$\pi = 9{,}640$$

This operation is making money on H_y; $A \neq 0$ is costly, but necessary as a breeding stock to maintain recruitment.

Other interesting solutions are possible. Suppose that Y were subsidized: $c_y \to -20$, expressing a value in having young fish in the population. With this change,

$$\pi = 20.2Y + 26A \qquad (3.61)$$

The optimum would lie at point C, where $H_y = 0$ and S is again at its maximum. Solution at this point is

$$Y = 893$$
$$A = 54$$
$$H_y = 0 \qquad (3.62)$$
$$H_a = 346$$
$$S = 1{,}000$$
$$\pi = 19{,}429$$

This subsidy policy is successful, in that Y has increased; contributing to that is H_y having gone to zero. The operation is making money on Y (subsidized) and H_a.

If instead, we make $p_y = 0$, in a different attempt to subsidize the young population, we get

$$\pi = -2Y - 4A \qquad (3.63)$$

Under this permutation, point B is preferred; there is no harvest and no population. There is no money to be made, and the optimal decision is to fold the business rather than lose money. The policy fails in this case, as it results in no population at all.

"Doing nothing" is the sustainable steady state preferred in this economy.

$$Y = 0$$
$$A = 0$$
$$H_y = 0 \qquad (3.64)$$
$$H_a = 0$$
$$S = 0$$
$$\pi = 0$$

The program **FishYAlp.xls** optimizes this problem. The optimization method is Linear Programming, as implemented in the Excel Solver.

3.4.2 Nonlinear Recruitment

In the previous problem, other constraints on resource availability could be added in addition to space. These would (potentially) further constrict the feasible region. The resource constraints define the carrying capacity here, as the linear, endogenous recruitment would drive the population to infinity without them.

One interesting limit is that imposed by nonlinear recruitment. As described above, consider the example function

$$R(A) = R_0(1 - e^{-A/A_0}) \qquad (3.65)$$

At low A/A_0, we have the linear approximation

$$R \simeq R_0 \frac{A}{A_0} \qquad (3.66)$$

and thus the correspondence

$$e(1 - \mu_e) \simeq \frac{R_0}{A_0} \qquad (3.67)$$

At $A/A_0 > 3$ or 4, R saturates at R_0. That saturation enforces a limit on the population itself, as we have seen in the earlier population dynamics section.

Developing the optimization problem as before, we have only to incorporate $R(A)$:

$$H_y = -(g + \mu_y)Y + R(A) \geq 0$$
$$H_a = gY - \mu_a A \geq 0$$
$$S = \sigma_y Y + \sigma_a A \leq \overline{S} \qquad (3.68)$$
$$Y \geq 0$$
$$A \geq 0$$

The constraint set is sketched in Figure 3.12, in the limiting case where space does not constrain ($\overline{S} \to \infty$). The curvature in the R function is clearly visible. Even in

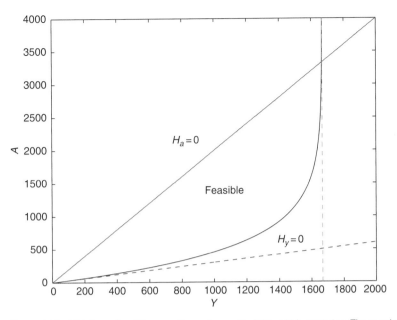

Figure 3.12. Fish farm: feasible operating options with R(A), which saturates. The constraints that $H_y \geq 0$ are shown as dotted lines for the two limiting approximations, low and high A. The constraint $H_a \geq 0$ is unaffected by the nonlinearity. Parameters as in the linear example; $R_0 = 1,000$; $A_0 = 500$. Note that at low A this corresponds to $e(1 - \mu_e) = 2$, lower than in the linear example, for plotting purposes.

this limiting case, it is clear that the saturation of R closes the feasible space, ruling out large solutions as beyond the carrying capacity of the system. The two limiting approximations for R (linear at low abundance and saturation at high abundance) are plotted as well. In this case, these provide an outside bound on the true nonlinear constraint with $R(A)$.

The objective is to maximize π as above:

$$\pi = p_y H_y + p_a H_a - c_y Y - c_a A$$
$$= p_y \left[-(g + \mu_y)Y + R(A)\right] + p_a \left[gY - \mu_a A\right] - c_y Y - c_a A$$
$$= \left[-p_y(g + \mu_y) + p_a g - c_y\right] Y + p_y R(A) + \left[-p_a \mu_a - c_a\right] A \qquad (3.69)$$

This result is analogous to Equation 3.57, with the nonlinear recruitment recognized.

The program **FishYAnLP.xls** optimizes this fish farm, retaining the previous linear example but adding the nonlinear recruitment function.

3.5 N STAGES

The formulations explored above are readily generalized to many interesting populations involving a larger set of stages: in order to resolve basic biological morphology and processes or to resolve management options relative to harvesting and recruitment. The basic structure is introduced in Section 3.1.2, with {B} the vector of biomass by stage:

$${B}^{k+1} = [L] {B}^k \tag{3.70}$$

The Leslie matrix ${L}$ here would have all processes internalized, including endogenous recruitment and harvesting. By resolving several stages, it is possible to represent recruitment into the youngest stage from all stages at reproductive maturity. Similarly, harvesting could be represented as directed at several stages, with multiple bycatch effects.

If recruitment were to be treated as exogenous, then that process would be removed from the Leslie matrix, leaving the formulation

$${B}^{k+1} = [L'] {B}^k + {R}^k \tag{3.71}$$

In both of the above, it is assumed that harvesting ${H}$ can be described with fishing mortality parameters μ_f embedded in $[L]$. In cases where the harvest itself can be controlled, then isolating this to an exogenous process makes sense:

$${B}^{k+1} = [L''] {B}^k + {R}^k - {H}^k \tag{3.72}$$

Finally, it is useful in some contexts to externalize the natural mortality to recognize the effects of other populations and/or the ecosystem generally:

$${B}^{k+1} = [L'''] {B}^k + {R}^k - {H}^k - {\mu}^k \tag{3.73}$$

All the dynamical contexts developed above are applicable: the eigenvalue/vector analysis for stable population modes; the general transient; the steady state and the approach to it; and the optimization of ${R}$ and/or ${H}$.

3.6 RECAP

This chapter is a continuation of Chapter 2. It adds further biological realism to the depiction of the living resource; the harvesting regime described is generally the same, with simple adjustments. Notably, the "fishing mortality" here is a composite of effort and technology, described as separate (multiplicative) factors in Chapter 2. Important ideas include the Leslie matrix representation of vital rates and the associated eigenproblem. Such a linear system either collapses or grows without bound. For many realistic populations, the latter would occur if not for other factors limiting population size. *Recruitment* is studied here as limiting at high abundance, reflecting exogenous controls on it. An extension illustrates an alternate harvesting regime more indicative of domesticated populations, rather than wild ones. In this case, the harvest resembles a management decision, whereas wild harvesting requires basic decisions about effort.

This is the beginnng of many possible enrichments of the biological representation. The reader is directed to Getz and Haight [27], Allman and Rhodes [1], Murray [65], Britton [6], Quinn and Deriso [74], Caswell [8], Cushing [16], Hoppensteadt [41], Renshaw [75], and Fennel and Neumann [23]. This remains an active and productive area of research.

3.7 PROGRAMS

The following programs illustrate the ideas in these lectures:

FishYA.xls (ex2box.m) illustrates the two-stage population and its Leslie matrix dynamics.
FishYA_R1.xls adds autocorrelated random distrubance to **FishYA**.
FishYAlp.xls optimizes the two-stage fish farm with space limitation, via LP as implemented in the Excel Solver.
FishYAnlp.xls optimizes the two-stage fish farm with infinite space, but nonlinear recruitment.
Fish-3-box.xls simulates a three-stage population (Problem 2), with nonlinear recruitment.
Fish-3-box_rand.xls adds autocorrelated random disturbances to **Fish-3-box**.
Fish-3-box_LP.xls optimizes a three-stage fish farm.

3.8 PROBLEMS

1 Population structure of a certain fish is described by a two-stage model accounting for juveniles and adults. Adults spawn once a year, producing 100 eggs per adult. Egg mortality is 80%. Surviving eggs become juveniles. Average age of maturity to adulthood is four years, during which mortality is negligible. Subsequently, adults live an additional 10 years on average, before dying a natural death. These data pertain to the unharvested population.

Fishing effort is directed at the adults. There is a bycatch in which juvenile fishing mortality is one-half of adult fishing mortality. Rules are in force that bycatch must be thrown back; so the juveniles caught die, but they are not part of the marketable harvest.

Adult fishing mortality is 30% per year.
(a) Develop the Leslie matrix for this situation.
(b) Find the growth rate of the population and its mode (ratio of juveniles to adults).
(c) At high population levels, recruitment is limited by exogenous factors to 400,000 surviving eggs per year. Find the steady-state population of juveniles and adults, as well as the steady-state harvest.
(d) Eliminate the bycatch and re-solve (c). What is the effect on the harvest and why?

(e) The fishery is in the steady state from (c). It is decided to eliminate the bycatch and operate at the steady state (d). At what rate will the population approach this new steady state?

2 Consider the following population: Adults spawn once per year, producing 20 eggs per year per adult, of which one-half survive to become young fish. After two years, these become adolescents, and after two more years they become mature adults. Natural mortality is 20% per year for young fish, 15% for adolescents, and 10% for adults. Fishing is directed at the adults, at 10% per year, with a bycatch effect of one-half (mortality = one-half of 10%) on the adolescents, which must be discarded when caught.

(a) Describe this population with a three-box, metered model using a one-year time step.

(b) Identify the stable population mode and its rate of growth.

(c) At high population levels, crowding in a limited-area habitat creates an upper limit on recruitment R_0. Express the steady-state population (number of individuals in each stage) as a function of R_0. Evaluate the same for $R_0 = 1,000$.

(d) For $R_0 = 1,000$, plot the steady-state harvest as a function of fishing mortality. More generally, recruitment R depends on the adult population A:

$$R = R_0(1 - e^{-A/100}) \tag{3.74}$$

(e) Simulate the growth of this population, starting with an initial spawning stock $A = 10$ and continuing to a steady state.
  (i) How long does the population take to stabilize?
  (ii) Is the steady state predicted above realized?
  (iii) Is the approach to steady state monotone, oscillatory, or chaotic?

(f) Add a stochastic disturbance to recruitment: $R = R_0(1 - e^{-A/100} + \epsilon)$, with ϵ random with zero mean, standard deviation σ, and autocorrelation ρ. Consider $\sigma = .25, .5, .75$; and $\rho = .5, .9, .95$. Is the harvest stochasically sustainable? If not, what management rules can improve things?

3 Show that Equations 3.28 and 3.43 are equivalent.

4 A marine population is characterized by three stages:
  Young fish: Ages 1 through 3; natural mortality is 20% per year
  Adult fish: Ages 4 through 9; natural mortality is 5% per year
  Senior fish: Ages 10 and above; natural mortality is 10% per year
Only the adult fish reproduce; recruitment is at the rate of 20 eggs per adult per year, with 2% egg survivorship. Senior fish do not reproduce. It is assumed that a fishing policy will allow harvesting of adults and seniors at the rate μ_f. Due to the practicalities, bycatch of young fish at the rate $.3\mu_f$ is unavoidable; those fish die but must be discarded at sea.

(a) Formulate a Leslie matrix representation of this system.

(b) Find the stable population mode and its growth rate for $\mu_f = 0$.

(c) Find the stable population mode and its growth rate for $\mu_f = .4$.

(d) Find the steady-state harvest, assuming recruitment saturates at the value R_0 for high biomass abundance.

5 Use the same population as in Problem 4 (a) through (c), except recruitment R is now thought to be independent of the adult population, at $R = 10{,}000$ per year and constant.

(a) Obtain expressions for young, adult, and senior fish in steady state and the steady-state harvest, as functions of μ_f.

(b) Plot the results of (a) versus μ_f.

(c) Show the Leslie matrix representation of this system.

(d) At what rate will the population approach the steady state just described?

6 A certain marine species spawns once a year. Each mature adult produces 50 eggs yearly, one-half of which survive and become juvenile fish. After three years, juveniles become reproductively mature adults. Natural mortality is 20% per year for juveniles, 10% per year for adults. Because population is at a very low level, fishing is prohibited. It will be allowed only if/when the population grows to 10 times its present size.

(a) Derive the Leslie matrix for this population.

(b) Find the stable population mode and its rate of growth.

(c) How long will it be before fishing may begin?

7 A population of fish produces 30 eggs per adult, once a year; two-thirds of the eggs survive to become young (prereproductive) fish. The young mature to adults after five years. Young mortality is 50% per year; adult mortality is 10% per year. There is no harvesting.

(a) Find the stable population mode and its rate of growth.

(b) Estimate the time required for this population to grow to 50 times its present size.

(c) Add adult fishing mortality at 50% per year, with a bycatch effect of one-fifth on the young fish (young fishing mortality is 10% per year). Find the stable population mode and its growth rate under these conditions.

(d) At high population levels, two things are different:

(i) Recruitment is limited to the constant value $R = 1,000$, independent of the number of fish.

(ii) Fishing mortality is 30% (adult) and 15% (juvenile, bycatch).

Express the steady-state value of young fish, adult fish, and legal (adult) harvest. Also find the illegal (juvenile) harvest.

8 Use the same population as in Problem 7, except that at high population levels recruitment is limited to the constant value $R = 1{,}000$, independent of the number of fish. Also, fishing mortality is real but unknown.

(a) Adjust the Leslie matrix representation from Problem 7 accordingly.

(b) Express the steady-state value of young and adult fish as functions of fishing mortality μ_f. Include the one-fifth bycatch effect.

   (c) It is observed that the steady-state population has the proportion two adults to three young fish. What is the value of fishing mortality?

9 Reconsider the population described in Problem 2. Formulate it subject to limited habitat as in a fish farm, with endogenous reproduction at the low population linear limit:

$$R = \frac{R_0}{100}A \qquad (3.75)$$

$R_0 = 1,000$. Use the annual harvests (not the rates μ) as the decision variables. Determine, via linear programming, the optimal yearly harvest in each stage. Use these parameters:

    Net profit per fish harvested: 10, 20, 10
    Cost of maintenance per fish: 1, 2, 2
    Habitat requirement per fish: 1, 4, 4
    Total available habitat: 5,000

10 Simulate the system of Problem 9, using the optimal harvesting policy.
   (a) Do the LP results work?
   (b) Add stochastic disturbances as in Problem 2-f. Formulate some management rules for unexpected situations.

11 Reconsider the optimization in Problem 9. Remove the habitat limit, but add non-linear (saturating) recruitment. Using nonlinear optimization, obtain and plot the optimal steady harvest versus price. Interpret.

12 Use the same population as in Problem 7. This fishery is to be operated in steady state by taking a fixed harvest H_y and H_a every year. There is a fish food limitation of 1,000 units of food per year. The following data apply to young and adult fish, respectively:
    Value of a harvested fish: 10, 15
    Annual cost of maintaining a fish in the system: 5, 5
    Annual food requirement per fish: 2, 1
   (a) Formulate this problem in steady state, as a linear programming problem.
   (b) Find the optimal solution graphically.
   (c) What are the optimal values of Y, A, H_y, H_a?

13 A small population of wild turkeys has been introduced to a region. Each adult pair reproduces annually, with 24 eggs per female, 75% of which survive. Juveniles take two years to reach adulthood. Natural mortality is 20% per year for juveniles, 10% per year for adults.
Note: Assume that one-half of all birds are female and account for all adults in "breeding pairs."
   (a) Describe the dynamics of this population with a Leslie matrix.
   (b) What is the stable population mode, and what is its growth rate?
   (c) Hunting is allowed, directed at adults only; mortality is increased by the rate μ_h. What value of μ_h will result in extinction of this species? (Hint: Redo Question (b) with hunting.)

(d) Assume no hunting. At high population levels, recruitment is limited to 1,000 juveniles per year. What is the steady-state population of juveniles? of adults?

14 An invasive species of marine fish exhibits two stages: juvenile and reproductive adults. Juveniles take four years on average to mature. Adults produce 20 surviving recruits each year; they die of old age on average after five adult years. There are no predators. All numbers refer to the female half of the population.
   (a) Describe the population dynamics in terms of a Leslie matrix.
   (b) What is the stable population mode, and what is its growth rate?
   (c) An eradication program is intended to eliminate all recruitment. How fast will the population decline?
   (d) The same as (c), but the eradication is only 90% effective. Will the program still work? What will be the rate of growth and/or decline?

15 A population has three stages: juvenile, adult female, and adult male. Sexual differentiation occurs at age 2; at that point, 40% become male, 60% female. Juvenile mortality is 20% per year; male mortality is 60%; and female mortality is 20%. Recruitment occurs at the exogenous rate of 1,000 juveniles per year. Fishing mortality is directed at all adults, at the rate of 30% per year. Males weigh 5 pounds each; females, 4 pounds each. There is a juvenile bycatch of 15% per year, which is fatal and illegal to sell.
   (a) What is the steady-state harvest in pounds of fish per year?
   (b) It is proposed to eliminate the bycatch. What is the steady-state harvest now?

16 Reconsider the fish farm example in Section 3.4.1.
   (a) Everything is known except c_y and c_a. Under what conditions will point A be optimal? point C? Express your answers as inequalities.
   (b) Ingore (a). Change g from 0.4 to 0.6 and keep all else fixed. Describe the new points, A and C.

17 Revisit the fish farm as formulated in Equations 3.50 through 3.57. Eliminate the endogenous recruitment; instead, make recruitment exogenous and constant at $R_0 = 400$.
   (a) List the constraints.
   (b) Revise Figure 3.11 accordingly.
   (c) Discuss the possible optimal solutions without knowing the economic parameters. How many interesting optima are there, and what characterizes each?

18 Continuation of Problem 17. Suppose R represents recruits bred in a separate nursery and bought at a cost of c_R per recruit. R can have any positive value below R_0. Formulate this as a linear-programming problem. Be clear about
   (a) the objective to be optimized
   (b) all the constraints

19 Derive Equation 3.38 for the closed-access equilibrium where π is maximized.

20 In Section 3.2.1, rent is expressed as a function of μ_f.

  (a) For the open-access case, find formulas for the steady-state values of A, Y, H, μ_f, and π.

  (b) Repeat for the closed-access case.

21 Following the text development at Equation 3.40,

  (a) Derive the expression for rent as a function of adult biomass, $\pi(A)$ (Equation 3.41).

  (b) Compare with the comparable expression for $\pi(B)$ in the simpler logistic case described in Chapter 2.

  (c) Using the result of (a), find the equilibrium A for open and for closed-access fisheries. Compare with those in the text or those found in Problem 20, where equivalently $\pi = \pi(\mu_f)$.

4 The Cohort

Here we present the final development of the living resource. The "cohort" imagined is a group of identical individuals. It is recruited into a population and develops thereafter, subject to natural and managed mortality. It is described in terms of number of individuals and the weight of the average individual. Once recruited, this cohort decreases in number while increasing in individual biomass. A basic decision on harvesting is twofold: when to begin, and how hard to work at it.

This analysis treats recruitment of each cohort as exogenous. It occurs regularly, timed with the population, *without* relation to the mature individuals surviving in previous cohorts. This is one of the limits studied in Chapter 3. In the fisheries literature, this corresponds to abundant spawning but highly variable survivorship in the larval stage, prerecruitment. In other contexts, it corresponds to completely controllable recruitment as in agriculture or silviculture, where cohorts are imported and "planted." Readers will recognize the Forester and Faustman criteria from the forestry literature here.

Cohorts are assumed to be long-lived (many years) relative to the recruitment interval (e.g., annual) and mixed together. Individual age is assumed identifiable; hence the notion of "year class" being the survivors of a certain recruitment event. Even when recruitment is uncontrollable, it is observable after the fact in the early-year classes.

Uncontrolled recruitment (e.g., wild fisheries) creates a challenge for monitoring and managing the population, and several distinct cases are explored. Perfectly controllable recruitment (e.g., managed forestry; aquaculture) creates the opposite challenge: deciding what level to set recruitment at.

This chapter concludes with a brief extension to the individual-based cohort.

4.1 SINGLE COHORT DEVELOPMENT

We describe a cohort of identical individuals as a function of age. The description assumes a one-time recruitment event, followed by growth, mortality, and harvesting of that cohort as it ages. We will start with a continuous description of the cohort with time (age), rather than discrete stages, as used in the previous chapter. As the problems get more complex, we can always fall back to a finite number of discrete stages and algebra rather than calculus.

4.1.1 Vital Rates

We start with the classic Beverton–Holt cohort model. Nomenclature used here is

t: age past recruitment
$W(t)$: the weight of an individual at age t
$N(t)$: the number of individuals at age t
R: recruitment at the start: $N(0) = R$
$B(t) = N(t) \cdot W(t)$: the biomass at cohort age t
μ: natural mortality, assumed constant
μ_f: fishing mortality
H: the harvest, accumulated over time
h: the harvest per recruit, $h = H/R$

The dynamics involve first-order (linear) rates, with no recruitment beyond the initial event. First, there is the intrinsic growth function of a surviving individual:

$$\frac{dW}{dt} = G \quad \text{(given)} \tag{4.1}$$

and the survival of the population is subject to natural mortality and harvesting (fishing):

$$\frac{dN}{dt} = -\mu N \quad t < t_f$$
$$\frac{dN}{dt} = -(\mu + \mu_f)N \quad t \geq t_f \tag{4.2}$$

The growth function $G(t)$ is given, as is the natural mortality μ. (Notice that these two features are separated here. G is *individual* growth of survivors; all mortality, including that due to harvesting, is accumulated in N, the number of survivors.)

There is a *two-parameter harvesting policy* (t_f, μ_f) describing (a) the minimun age at harvest, t_f; and (b) the harvest rate μ_f for individuals beyond that age. It is common to associate t_f with "net size," which ideally harvests only fish whose age exceeds t_f. The harvest accumulates over time:

$$\frac{dH}{dt} = 0 \quad t < t_f$$
$$\frac{dH}{dt} = \mu_f B = \mu_f WN \quad t \geq t_f \tag{4.3}$$

In the absence of fishing, the biomass as a function of age is

$$N(t) = Re^{-\mu t} \tag{4.4}$$

$$B(t) = RW(t)e^{-\mu t} \tag{4.5}$$

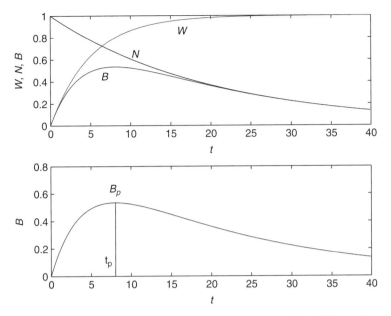

Figure 4.1. Cohort relations: individual weight W, survivorship N, and biomass B versus age t, for $R = 1$. There is zero fishing, $\mu_f = 0$. The biomass peak B_p occurs at age t_p.

It is assumed that natural growth in indiviudal mass tapers off in old age, eventually losing ground to natural mortality. The biomass would thus be expected to rise early with individual growth, peak, and decline in old age due to natural mortality. The rate is

$$\frac{dB}{dt} = R\left(\frac{dW}{dt} - \mu W\right)e^{-\mu t} \tag{4.6}$$

and we have a peak when

$$\frac{dW}{dt} = \mu W \tag{4.7}$$

or equivalently, when growth equals mortality. We identify this age as t_p. Beyond this age, the biomass decays due purely to natural processes. This is illustrated in Figure 4.1.

4.1.2 Fishing Mortality and Harvest

Fishing mortality is added beginning at age t_f in Figure 4.2. For this case, we have for $t > t_f$:

$$B(t) = R \cdot W(t)e^{-\mu t}e^{-\mu_f(t-t_f)}$$
$$= R \cdot W(t)e^{-\mu t_f}e^{-(\mu+\mu_f)(t-t_f)} \tag{4.8}$$

Higher μ_f increases the decay of the population once harvesting begins. The limit of infinite harvesting causes an instantaneous harvest and crash of B to zero.

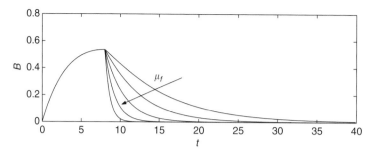

Figure 4.2. Biomass as in Figure 4.1, subject to fishing mortality μ_f beginning at t_f ($=t_p$ in this case). Increasing values of μ_f are shown.

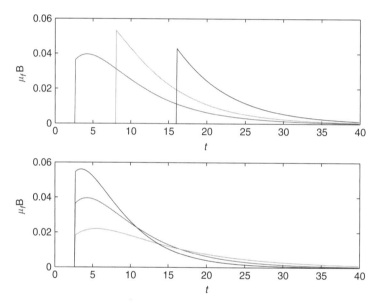

Figure 4.3. *Top:* Harvesting rate $\mu_f B$ starting at different t_f (before, at, and after t_p); identical μ_f. *Bottom:* Identical t_f; different μ_f. The intermediate curve in the bottom panel is the same as the early curve above.

Clearly, there is no benefit in beginning the harvest *after* t_p, as some harvest would be lost to natural mortality first. The simplest ideal harvest would have $(t, \mu_f) = (t_p, \infty)$, whereby the cohort would grow to its peak biomass B_p and then be entirely harvested to zero, instantly. Finite harvesting rates μ_f take time; in that case, there is incentive to begin *before* t_p, sacrificing some natural growth to avoid the effect of natural mortality later, during a protracted harvest.

The harvest rate is $\mu_f B$. It is illustrated in Figure 4.3. Different starting times clearly affect this, both in terms of the *onset* of harvesting and the *size* of the harvest. If the cohort is fully mature at t_f (W has reached its asymptote), then the harvest simply decays from initial value at the exponential rate $-(\mu + \mu_f)$; the total harvest is then proportional to the starting value of $\mu_f B$. Earlier harvesting is at the expense of natural growth. Figure 4.3 also illustrates the effect of μ_f on harvest rate, for a fixed t_f. Fast harvesting of mature fish gets ahead of natural mortality. Clearly, the harvesting history depends on t_f and μ_f.

The total harvest H accumulated over time is the integral of $\mu_f B$:

$$H = \mu_f R e^{-\mu t_f} \int_{t=t_f}^{t=\infty} W(t) e^{-(\mu+\mu_f)(t-t_f)} dt \qquad (4.9)$$

The accumulated *harvest per recruit*, $h \equiv H/R$, depends on the harvesting policy:

$$h = h(t_f, \mu_f) \qquad (4.10)$$

The *peak harvest per recruit*, h_p, is associated with instantaneous harvesting at t_p:

$$h_p = h(t_p, \mu_f \to \infty) = \frac{B_p}{R} \qquad (4.11)$$

Generally, we would expect harvest to begin earlier (t_f less than t_p) if the fishing rate is limited (μ_f small), for reasons mentioned above. By the same reasoning, we would expect t_f to approach $t_f = t_p$ for huge μ_f. The *Eumetric harvest h^** is achieved for any given μ_f, with the ideal t_f^*, which maximizes the accumulated harvest:

$$t_f^* = t_f^*(\mu_f) \qquad (4.12)$$

$$h^*(\mu_f) = h(t_f^*, \mu_f) \qquad (4.13)$$

Notice these are strictly *biological* ideas; there is no economy here yet.

4.1.3 Instantaneous Harvest: Uniform Annual Increment

The harvest averaged over time is h/t. For an *instantaneous harvest* at $t = t_f$, it is

$$\frac{h}{t_f} = \frac{1}{t_f} \frac{B}{R} \qquad (4.14)$$

This is maximal when

$$\frac{dB}{dt_f} = \frac{B}{t_f} \qquad (4.15)$$

Clearly, this will occur before B_p is reached (where $dB/dt_f = 0$). It is illustrated below in Figure 4.4. This is referred to as the *Forester's criterion* for harvest time t_f. It is sensible in the case where harvesting can be followed by replanting, thus repeating the cycle forever and providing a uniform annual yield as "rent" for the use of the ecosystem supporting the cohort over its lifetime. In this context, the average annual harvest $\frac{B}{t_f}$ is referred to as the *uniform annual increment*. The Forester's criterion is that the marginal biomass growth be equal to the average biomass growth.

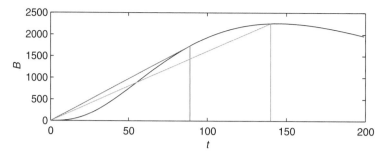

Figure 4.4. Instantaneous harvesting illustrated for harvest times $t_f = 90$ and 140. The uniform annual increment is B/t_f, graphically, the tangent of the line from the origin to $B(t_f)$. As in Equation 4.15, this is maximized when this line just touches the $B(t)$ graph – the Forester's criterion. Clearly, the Forester's criterion asks for harvesting before t_p. (See also Figure 4.10.)

4.2   EXAMPLE

Suppose we have the growth function

$$G(W) = g\left(\overline{W} - W\right) \tag{4.16}$$

which integrated from zero initial W, gives the weight function

$$W(t) = \overline{W}(1 - e^{-gt}) \tag{4.17}$$

From Equation 4.7, we have at peak,

$$G_p = \mu W_p \tag{4.18}$$

and for this growth function we get

$$W_p = \left(\frac{g}{\mu + g}\right)\overline{W} \tag{4.19}$$

W_p occurs at t_p; from Equation 4.17, we have

$$t_p = \frac{1}{g}\ln\left[\frac{g + \mu}{\mu}\right] \tag{4.20}$$

Substituting for $B = RW(t_p)e^{-\mu t_p}$, we have the peak biomass B_p:

$$B_p = R\overline{W}\left[\frac{g}{g + \mu}\right]\left[\frac{\mu}{g + \mu}\right]^{\mu/g} \tag{4.21}$$

and the peak harvest per recruit, $h_p = B_p/R$:

$$h_p = \overline{W}\left[\frac{g}{g + \mu}\right]\left[\frac{\mu}{g + \mu}\right]^{\mu/g} \tag{4.22}$$

The pair (t_p, h_p) assume an instantaneous harvest when biomass peaks; effectively, $\mu_f \to \infty$ at $t = t_p$.

More generally, we have finite μ_f. An exercise in integration of exponentials will yield, from Equation 4.9:

$$h(t_f, \mu_f) = \mu_f \overline{W} \left[\frac{1}{\mu + \mu_f} e^{-\mu t_f} - \frac{1}{g + \mu + \mu_f} e^{-(g+\mu)t_f} \right]$$

$$= \mu_f \overline{W} e^{-\mu t_f} \left[\frac{1}{\mu + \mu_f} - \frac{1}{g + \mu + \mu_f} e^{-g t_f} \right] \qquad (4.23)$$

Differentiating this with respect to t_f, while keeping μ_f constant, we find the maximum harvest of this cohort at $\partial h / \partial t_f = 0$. This defines the ideal minimum fishing age t_f^*:

$$t_f^* = \frac{1}{g} \ln \left(\left[\frac{\mu + \mu_f}{\mu} \right] \left[\frac{g + \mu}{g + \mu + \mu_f} \right] \right) \qquad (4.24)$$

Combining this with Equation 4.20 above, we get:

$$t_f^* = t_p - \frac{1}{g} \ln \left[\frac{g + \mu + \mu_f}{\mu + \mu_f} \right] \qquad (4.25)$$

All the rates are nonnegative, so we see that t_f^* is always $\leq t_p$, with the upper limit $t_f^* = t_p$ at infinite harvesting rate.

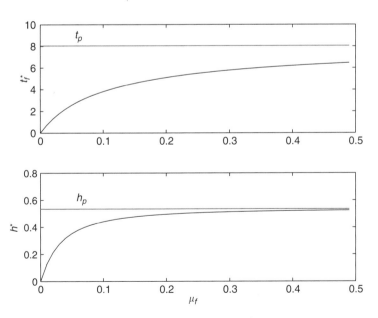

Figure 4.5. *Top:* Optimal harvest time t_f^* versus μ_f. t_f^* is bounded by the natural peak biomass age t_p as indicated. *Bottom:* Eumetric harvest per recruit h^* versus μ_f, bounded by h_p. Parameters: $(g, \mu, \overline{W}) = (0.2, 0.05, 1.0)$. These give the asymptotes shown: $t_p = 8.047$; $h_p = 0.535$.

The *Eumetric yield* $h^*(\mu_f)$ uses optimal t_f^* at any given μ_f:

$$h^*(\mu_f) = h(t_f^*, \mu_f) \tag{4.26}$$

For this example, $h(t_f, \mu_f)$ is given above in closed form (Equation 4.23). Figure 4.5 is a plot t_f^* and h^* vs. μ_f for a given set of parameters. The upper bound on these, as $\mu_f \to \infty$, is

$$t_f^* \to t_p$$
$$h^* \to \overline{W}e^{-\mu t_p}\left[1 - e^{-g t_p}\right] = B_p/R \equiv h_p \tag{4.27}$$

Suppose the harvesting were suboptimal – for example, if there were no control of t_f and effectively μ_f were applied from $t = 0$. Equation 4.23 governs here, with $t_f = 0$.

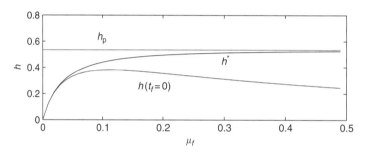

Figure 4.6. Harvest per recruit, versus μ_f, for the example. The suboptimal h holding $t_f = 0$ is plotted. Necessarily, this cannot exceed the Eumetric curve h^*, reproduced here from Figure 4.5. Parameters are as in Figure 4.5.

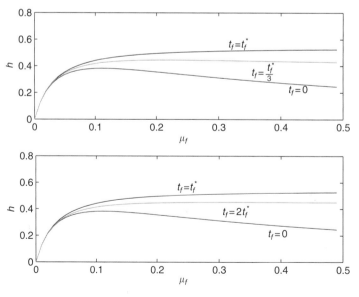

Figure 4.7. Harvest per recruit, versus μ_f, for the example. The suboptimal h holding $t_f = 0$, $t_f^*/3$, and $2t_f^*$ are plotted. Necessarily, these cannot exceed the Eumetric curve h^*, reproduced here from Figure 4.5. Parameters are as in Figure 4.5.

The results are plotted in Figure 4.6 for this example, versus μ_f. Clearly, the harvest is below the Eumetric limit and thus suboptimal. At low μ_f, the ideal t_f^* approaches zero, so the two curves converge there. At higher μ_f, it is useful to delay harvesting. Not doing so results in less overall harvest. Figure 4.7 adds the suboptimal harvest with $t_f = t_f^*/3$ and $t_f = 2t_f^*$. The latter curve confirms that t_f^* is in fact superior to either earlier or later harvesting.

The reader is encouraged to explore an alternate, logistic fromulation of $G(W)$ in the exercises (Problems 3 through 6).

4.3 ECONOMIC HARVESTING

The developments so far have focused only on biological development of a single cohort. The harvest described occurs over time according to the two-parameter policy (t_f, μ_f). The policy selection needs to reflect two important features: (a) the recruitment scenario; and (b) the economics of the situation.

There are several important cases:

1. One-time recruitment only. This would be the case of the "bonanza cohort." With no future recruitment to be relied upon and eventual death of the cohort, one would engage in "cohort mining," focusing on present worth. This is a *nonrenewable resource*, maturing and then dying.

2. Repeated recruitment at the same rate R each year. This is a *renewable resource*; a harvesting policy would produce the same *annual* yield, once the various year classes had achieved a steady presence. The harvesting would yield portions of all cohorts that had reached age t_f; all surviving cohorts would be harvested simultaneously. An idealized fishery with *dependable* annual recruitment would mix annual cohorts in this way. The *Eumetric harvest* could be realized in a single year, across many extant cohorts of different ages, instead of over time.

3. Sequential harvesting, followed by intentional reseeding (rerecruitment). This scenario would be relevant when recruitment could be arranged and when subsequent cohorts could not mutually coexist. Agricultural cropping is close to this, as is forestry. Recruitment would be managed, once the previous cohort was completely harvested and the land cleared. This leads to the standard Faustman model of economic harvesting.

4. Repeated annual recruitment but at a *variable*, uncontrolled rate. Here we have no standard guide except that provided in the statistics of R. One can imagine a harvesting policy based on mean annual recruitment; its standard deviation; either of its extremes; a nowcast/hindcast estimate of B and its age structure; and/or a forecast of R. This is generally beyond the scope of this text; see, for example, Mangel [59].

In each case, we will assume an annual cost of effort $C = c \cdot \mu_f$ as in previous chapters. Generally, we will have to consider the time value of money in terms of interest rate r.

4.3.1   Cohort Mining

Economic Lifetime

Here we have an isolated cohort; it will die if not harvested. We imagine a fixed harvest policy (t_f, μ_f) that is constant over time. We have the harvest rate times the price of the harvest p less the cost of the harvest. The rate of rent accumulation is

$$\dot{\pi} = p\mu_f B - c\mu_f = p\mu_f \left(B - \frac{c}{p} \right) \tag{4.28}$$

Harvesting is not profitable when B is too small, irrespective of μ_f. Specifically, profitability requires

$$B(t) \geq \frac{c}{p} \equiv B_0 \tag{4.29}$$

This break-even point[1] can be crossed twice: once during the early growth, and then later during the decline of B. In the growth stage, economic harvesting cannot begin until $B(t_f) \geq B_0$, and this sets the lower limit on t_f. Subsituting for $B(t)$ from Equation 4.5, we have

$$RW(t_f)e^{-\mu t_f} \geq B_0 \tag{4.30}$$

and therefore,

$$t_f \geq \frac{1}{\mu}\ln\left[\frac{RW(t_f)}{B_0} \right] \tag{4.31}$$

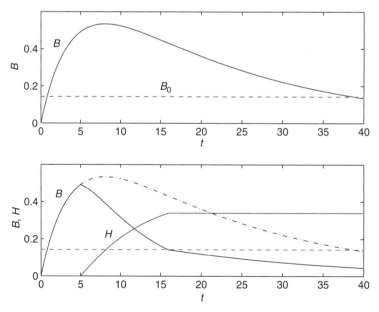

Figure 4.8. Mining of a single cohort. *Top:* Fishing is economically viable when $B > B_0 = c/p$. *Bottom:* evolution of B and accumulation of harvest H with fishing beginning at $t_f = 5$ and ending when $B < B_0$. The dotted lines are B for the no-fishing case and the unchanged limit B_0, reproduced from the top panel.

---

[1]   We encountered this value of B_0 in the previous chapter, scaled to reflect the different units of c.

Harvesting will be terminated economically if B is too small, such that $\dot{\pi} < 0$; that point is reached when

$$B(T) = \frac{c}{p} = B_0 \qquad (4.32)$$

This sets the cohort's economic lifetime T; profitable harvesting can occur for $t_f \leq t \leq T$. Subsituting for $B(t)$ from Equation 4.8, we have

$$RW(T)e^{-\mu T}e^{-\mu_f(T-t_f)} = B_0 \qquad (4.33)$$

and therefore,

$$T = \frac{1}{\mu + \mu_f}\left(\ln\left[\frac{RW(T)}{B_0}\right] + \mu_f t_f\right) \qquad (4.34)$$

In the practical case where W has reached full maturity $\overline{W}$, the asymptote is

$$T = \frac{1}{\mu + \mu_f}\left(\ln\left[\frac{R\overline{W}}{B_0}\right] + \mu_f t_f\right) \qquad (4.35)$$

Notice that the economic lifetime T scales as $\ln(R)$ in this limit and linearly with t_f:

$$\frac{\partial T}{\partial t_f} = \frac{\mu_f}{\mu_f + \mu} \qquad (4.36)$$

Instantaneous Harvest

Financially, the harvest needs to be discounted to present value. An easy limit can be reached fast: the *instantaneous harvest*, $\mu_f \to \infty$ and $T = t_f$:

$$\pi = p\left(B(t_f) - B_0\right)e^{-rt_f} \qquad (4.37)$$

Its maximum is at $\partial\pi/\partial t_f = 0$:

$$\frac{\partial\pi}{\partial t_f} = pe^{-rt_f}\left[\frac{\partial B}{\partial t} - r(B - B_0)\right]_{t=t_f} = 0 \qquad (4.38)$$

The optimum is simple:

$$\frac{\partial B}{\partial t} = r(B - B_0) \qquad (4.39)$$

Immediately, we see a simple effect of r. Harvesting occurs *while B is still growing* ($\partial B/\partial t > 0$), and therefore, $B < B_p$ and $t_f < t_p$. The economic harvester does not wait for B to peak; as soon as biological growth is less than financial growth, B is harvested and sold. Increasing r shortens t_f. The uneconomic remnant in any case is the same, B_0, which is left to natural mortality. This is sometimes referred to as the *Fisher criterion* [68]. When $r = 0$, this optimum reverts to a purely biological maximum at t_p, B_p, discussed before.

The instantaneous harvest H is $B - B_0$, so Equation 4.39 is restated:

$$\frac{\partial H}{\partial t} = rH \qquad (4.40)$$

The trade-off is between harvest growth "in the bank" versus harvest growth "in the wild."

In terms of individual weight W, we have Equations 4.5 and 4.6. Their substitution into Equation 4.39 gives

$$\frac{dW}{dt} = (\mu + r)W - r\frac{B_0}{Re^{-\mu t_f}} \tag{4.41}$$

When $r = 0$, we recover the biological maximum, Equation 4.7. Increasing r results in earlier (younger) harvesting; if it is costly, harvesting is delayed.

Extended Harvest

For *finite* μ_f, we can expect the same effect as noted from strictly biological considerations: Start early in order to get ahead of natural mortality. Thus, t_f will be lower than that for an instantaneous harvest.

Generally, π is the discounted present worth accumulation of $\dot{\pi}$:

$$\pi = p\mu_f \int_{t_f}^{T} (B - B_0)e^{-rt}dt \tag{4.42}$$

(As in Equation 4.28, $B_0 \equiv c/p$ here accounts for the cost of harvesting effort μ_f.) The present worth calculation adds the additional incentive to start early: Avoid the decay of present worth associated with delay.

We need to be careful in that for fixed μ_f, $B = B(t_f, t)$ in this integral. (See Equation 4.8.) Similarly, T depends on t_f (Equation 4.36). Keeping μ_f constant, the derivative of π is

$$\frac{\partial \pi}{\partial t_f} = p\mu_f \left[\int_{t_f}^{T} \frac{\partial B}{\partial t_f} e^{-rt}dt + (B - B_0)e^{-rt}\Big|_T \frac{\partial T}{\partial t_f} - (B - B_0)e^{-rt}\Big|_{t_f} \right] \tag{4.43}$$

From Equation 4.8, we have

$$\frac{\partial B}{\partial t_f} = \mu_f B \tag{4.44}$$

which is always positive. Equation 4.36 gives us positive $\partial T/\partial t_f$, but $(B(T) - B_0)$ is zero, so this term vanishes. Assembling these gives

$$\frac{\partial \pi}{\partial t_f} = p\mu_f \left[\int_{t_f}^{T} \mu_f B e^{-rt}dt - (B(t_f) - B_0)e^{-rt_f} \right] \tag{4.45}$$

so a condition for maximum π, at constant μ_f, is

$$\int_{t_f}^{T} \mu_f B e^{-rt}dt = (B(t_f) - B_0)e^{-rt_f} \tag{4.46}$$

4.3.2 Sustained Recruitment of Mixed Cohorts

Here we imagine an ecosystem that recruits a multiyear cohort every year and in which all cohorts are mixed without competition or crowding. The same harvesting policy (μ_f, t_f) operates on each simultaneously: μ_f takes a portion of the older ones, age $> t_f$, and ignores the younger ones. The harvest is simply summed across cohorts. Once a few cohort lifetimes have been recruited, the system is loaded in a steady state such that harvesting across cohorts in a single year is equivalent to whole-life harvesting of a single cohort. Accumulating across cohorts is the same as accumulating across time for a single cohort.

We thus have a system that will return, in a single year, the whole-life harvest defined for a single cohort. This would be a biological annuity without any further financial discounting, to be weighted against the cost of the fishing effort μ_f. Immediately, we see the inspiration for the Eumetric yield curve, $h^*(\mu_f)$ (Equation 4.13) in the context of cross-cohort harvesting. For any μ_f, the Eumetric t_f^* provides maximal harvest h^*. (See the example in Figure 4.6.)

The annuity $\dot{\pi}$ in this case is the value of the harvest less the cost of the effort:

$$\dot{\pi} = pRh^*(\mu_f) - c\mu_f \qquad (4.47)$$

This is plotted in Figure 4.9. The example is from Figure 4.6, comparing Eumetric harvesting with optimal $t_f(\mu_f)$ versus suboptimal harvesting with no control on

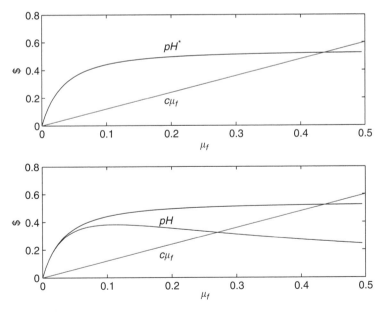

Figure 4.9. Annual rent versus effort for mixed cohorts. *Top:* Eumetric harvest $H^*(\mu_f)$. The vertical distance between value and cost is $\dot{\pi}$. There is a broad range of rent-producing options μ_f. *Bottom:* Nonoptimal H. The Eumetric harvest line is reproduced from the top panel. This suboptimal harvest policy restricts harvest, rent, and the profitable range of μ_f. Harvest curves are as in Figure 4.6.

t_f. In either case, there is a rent-maximizing point at intermediate effort μ_f and a rent-dissipating point at high effort. Clearly, the suboptimal case restricts the rent-producing options.

This brings us back, with a more realistic biological cohort orientation, to material introduced earlier in Chapter 2-style surplus-yield analysis of effort, harvest, biomass, and rent. All of that analysis is immediately relevant as the bioeconomic extension of this cohort case.

4.3.3   Sequential Cohorts: The Faustman Rotation

Here we examine sequential harvesting, followed by intentional reseeding (rerecruitment). This scenario would be relevant when recruitment could be arranged and when subsequent cohorts could not mutually coexist. Agricultural cropping and forestry are close to this. Recruitment would be managed, once the previous cohort was completely harvested and the land or ecosystem cleared. This leads to the standard Faustman model of economic harvesting.

Instantaneous, costless harvest followed by immediate reseeding produces an infinite series of identical harvests H separated by their maturation period t_f. In terms of the value p of the biomass, we have the present worth:[2]

$$\pi = p \left[\frac{B(t_f)}{e^{rt_f}} + \frac{B(t_f)}{e^{2rt_f}} + \frac{B(t_f)}{e^{3rt_f}} + \cdots \right] \tag{4.48}$$

$$= p \frac{B(t_f)}{e^{rt_f}} \left[1 + e^{-rt_f} + e^{-2rt_f} + e^{-3rt_f} + \cdots \right] \tag{4.49}$$

$$= p \frac{B(t_f)}{e^{rt_f} - 1} \tag{4.50}$$

Maximizing π gives the Faustman criterion:

$$\frac{dB}{dt_f} = rB \left[\frac{1}{1 - e^{-rt_f}} \right] \tag{4.51}$$

For long rotation periods t_f, this is essentially the same as the Fisher criterion developed above (Equations 4.39 and 4.40) for the present worth of a single cohort. (The criterion for "long life" is economic: large rt_f.) Here we add subsequent identical cohorts, to make a dependable annuity on a rotation period t_f. The effect of the subsequent cohorts is to increase the apparent r by the factor $\left[1/(1 - e^{-rt_f}) \right]$ relative to the Fisher criterion. The effect is not major for long-lived resources ($rt_f = 3$ or more), but it leads systematically to earlier harvesting. Essentially, the incentive is to harvest a little sooner so that the next cohort can get going. But in this long-lived case, the lion's share of the rent is realized in the very first harvest.

---

[2]   Here the identity $1 + x + x^2 + \cdots = \frac{1}{1-x}$ is useful, for $x^2 < 1$.

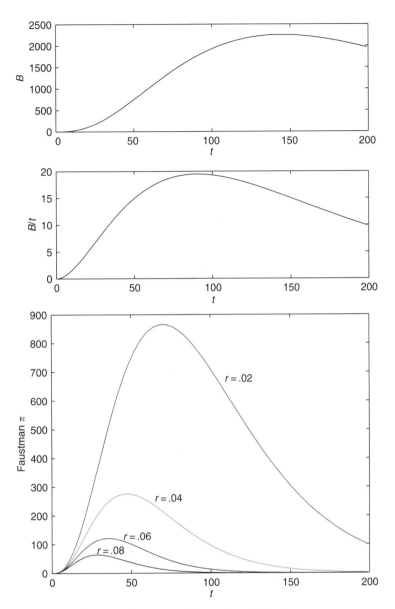

Figure 4.10. Biomass, uniform annual increment, and Faustman π versus age, as a function of interest rate r. Assumed is instantaneous harvest followed by rerecruitment, repeating forever. The example is as given in Equations 4.54 and 4.55; the vital rates are identical to those illustrated in Figure 4.4. *Top:* Biomass versus time, peaking arount $t = 140$. *Middle:* Uniform annual increment versus age; this peaks while B is still rising, around $t = 90$. *Bottom:* Rent based on costless harvesting, as a function of r and harvest age. Interest accelerates the harvest and reduces rent. Ultimately, at high r or slow growth, π from the living cohort is not competitive with other uses – for example, very short-lived cohorts or nonliving uses of the land (parking lot). (See also Figure 4.4.)

Above we have the net present value of the harvest; if we add the cost of reseeding or recruitment c, then

$$\pi = -c + \frac{pB - c}{e^{rt_f}}\left[1 + e^{-rt_f} + e^{-2rt_f} + e^{-3rt_f} + \cdots\right]$$

$$= -c + \frac{pB - c}{e^{rt_f} - 1} \tag{4.52}$$

leading to the optimum

$$\frac{dB}{dt_f} = r\left(B - \frac{c}{p}\right)\left[\frac{1}{1 - e^{-rt_f}}\right] \tag{4.53}$$

It is easy to interpret this value of π (Equation 4.52) as the financial value (present worth) of all future ecosystem output under managed rotation; and the Faustman criterion (Equation 4.53) as the rotation period t_p that maximizes π.

If π is the present value of an infinite series, then its sustainable annual worth is the annuity equivalent, $r\pi$, the *Faustman annuity*. It is evident that the optimal rotation period for π and $r\pi$ is the same.

In the context suggested, the enormous effect of r on ecosystem value π and on rotation length t_f should not be overlooked:

- Increasing r drives both π and t_f down for a given species.
- Increasing r causes a myopic focus on the first cohort only, in keeping with the present-value orientation of the Faustman criterion.
- Increasing r encourages substitution of better rent-producing species: shorter-lived, faster, like money growing at rate r.
- Old growth: When such a cohort is discovered (e.g., a 200-year-old forest), then it is an *economically* nonrenewable resource; the cohort will be mined and replanted with a Faustman rotation of either faster species and/or inorganic uses.

As an example, consider the vital rates

$$W(t) = t^2(1 - e^{-gt}) \tag{4.54}$$

$$N(t) = Re^{-\mu t} \tag{4.55}$$

These relations were used to plot Figure 4.4 to illustrate the uniform annual increment. Figure 4.10 illustrates these points among growth and harvesting according to the Forester and Faustman criteria. The harvest and the rent are simply plotted versus harvest age; the relevant optimal age is evident in the peaks of the curves. The student is encouraged to compute them from the marginal conditions cited, and to supplement figure 4.10 with the Faustman annuity $r\pi$.

4.4 UNCONTROLLED RECRUITMENT

4.4.1 Biomass

Here we imagine a sequence of cohorts, each initiated by a discrete, impulsive recruitment event on an ecologically determined clock. Let $(t_1, t_2, \ldots, t_i, \ldots)$ be the recruitment date and $(R_1, R_2, \ldots, R_i, \ldots)$ be the number of age-0 recruits in the ith cohort. All individuals have the same vital rates, in particular the natural mortality rate μ and the weight-vs.-age function W. The cohort growth relations preharvest are as in Equations 4.4 and 4.5, simply adjusted for cohort age $(t - t_i)$:

$$N_i(t) = R_i e^{-\mu(t-t_i)} \tag{4.56}$$

$$B_i(t) = R_i W(t - t_i) e^{-\mu(t-t_i)} \quad (t - t_i < t_f) \tag{4.57}$$

Harvesting begins at age t_f for each cohort, and the harvest policy (t_f, μ_f) is assumed uniform across all cohorts. From Equation 4.8, we have for the ith cohort

$$B_i(t) = R_i \cdot W(t - t_i) e^{-\mu t_f} e^{-(\mu+\mu_f)(t-t_i-t_f)} \quad (t - t_i > t_f) \tag{4.58}$$

Here it is convenient to define *recruitment to fishable age* $\mathcal{R}_i$ as the *number of indiviudals surviving to* t_f:

$$\mathcal{R}_i \equiv R_i e^{-\mu t_f} \tag{4.59}$$

Thus, we have in the harvestable population the sum of all cohorts age t_f and above:

$$\beta(t) = \sum_{t-t_i>t_f} \mathcal{R}_i \cdot W(t - t_i) e^{-(\mu+\mu_f)(t-t_i-t_f)} \tag{4.60}$$

β accounts for the total *fishable* population. The harvest is directed at $\beta(t)$ with uniform μ_f across all age-appropriate cohorts. β is supplemented by a nursery population of cohorts with age $< t_f$ – by policy, too young to harvest but still part of the total biomass B.[3]

Dynamically, we conceive of $\beta(t)$ (and/or $B(t)$) as a continuous evolution of a sum of extant cohorts, driven by discrete impulsive recruitment of subsequent cohorts. R_i is unpredictable, but its timing is regular. The harvest policy dictates a mandatory

---

[3] Aside: Recruitment to fishable age $\mathcal{R}_i$ indicates number of individuals. It may be defined alternatively, in terms of biomass, by simply invoking the individual weight function W:

$$\mathcal{B}_i \equiv R_i e^{-\mu t_f} W(t_f) = \mathcal{R}_i W(t_f) \tag{4.61}$$

$\mathcal{B}_i$ expresses the initial number in cohort R, subject to natural mortality *plus* individual development, up to the point t_f when the cohort enters the fishable biomass. In terms of $\mathcal{B}_i$, the fishable biomass in Equation 4.60 is reexpressed as

$$\beta(t) = \sum_{t-t_i>t_f} \mathcal{B}_i \cdot w(t - t_i) e^{-(\mu+\mu_f)(t-t_i-t_f)} \tag{4.62}$$

Here the individual weight function $w(t)$ is simply weight-at-age normalized by $W(t_f)$, the individual weight at fishable age: $w(t) = W(t)/W(t_f)$. This approach has its own special appeal, especially in cases where W reaches saturation or in cases of instantaneous harvest, $\mu_f \to \infty$.

delay t_f between R and $\mathcal{R}$ and a charactistic time of cohort persistence beyond t_f, given as $\frac{1}{\mu+\mu_f}$. Cohorts are inseparably mixed and concurrently harvested by the single harvesting policy (μ_f, t_f). Cohort members may be identifiable by age at capture.

4.4.2   Harvest

The harvest is continuous, summed over all fishable cohorts, at the rate μ_f. It is already accounted for in each cohort's biomass above (Equation 4.60). It is useful to integrate it over a single period (year, Δt) between recruitment events:

$$\frac{dH_i}{dt} = \mu_f \beta_i \tag{4.63}$$

$$\Delta H_i(t_0) = \mu_f \int_{t_0}^{t_0+\Delta t} \beta_i dt \tag{4.64}$$

and the total harvest in Δt is the sum over all fishable cohorts:

$$\Delta H(t_0) = \mu_f \int_{t_0}^{t_0+\Delta t} \beta dt \tag{4.65}$$

$$= \mu_f \sum_i \mathcal{R}_i \int_{t_0}^{t_0+\Delta t} W(t-t_i) e^{-(\mu+\mu_f)(t-t_i-t_f)} dt \tag{4.66}$$

If we make the harvest year start immediately after the most recent recruitment event, then we can improve the precision of the limits. For a harvest beginning at t_0, the youngest fishable cohort K is delayed by t_f:

$$t_K = t_0 - t_f \tag{4.67}$$

This and all older cohorts contribute to the harvest.

$$\Delta H(t_0) = \mu_f \sum_{i=-\infty}^{K} \mathcal{R}_i \int_{t_0}^{t_0+\Delta t} W(t-t_i) e^{-(\mu+\mu_f)(t-t_i-t_f)} dt \tag{4.68}$$

But clearly, very old cohorts contribute effectively nothing here; and the lower summation limit (earliest t_i) can be made finite for practical purposes. Using the combined mortality time scale, the oldest practical cohort was recruited $L\Delta t$ prior to cohort K:

$$(\mu + \mu_f)(L\Delta t) \simeq 4 \tag{4.69}$$

In other words, the oldest practical cohort has been fished long enough that $e^{-4} \sim 2\%$ or less of its initial members are still surviving.

$$\Delta H(t_0) \simeq \mu_f \sum_{i=K-N}^{K} \mathcal{R}_i \int_{t_0}^{t_0+\Delta t} W(t-t_i) e^{-(\mu+\mu_f)(t-t_i-t_f)} dt \tag{4.70}$$

Here is a little recap of the various time conventions used here:

- t_i is the time of recruitment for cohort i.

- t_p is the time interval after recruitment when the unharvested cohort peaks in biomass.
- t_0 is the start time of the harvest ΔH; Δt is its duration.
- t_f is the minimum harvest age.
- $t_K = t_0 - t_f$ is the recruitment time for the youngest harvestable cohort.
- $t_i + t_f$ is the time when harvesting begins for cohort i.
- L is the cohort persistence in the annual harvest. Each cohort i survives for the duration $t_f + L\Delta t$ beyond recruitment t_i. At any time, the yearly harvest ΔH involves the youngest fishable cohort K plus its L immediate prececessors.
- B is the total biomass; β is that portion that has matured to age $\geq t_f$. Harvesting is directed at β.

4.4.3 ### Example

As an example, consider the weight function used earlier:

$$W(t) = \overline{W}(1 - e^{-gt}) \tag{4.71}$$

Here we will use the following parameters: $(g, \mu) = (.4, .05)$; $(t_f, \mu_f) = (3.0, .4)$; $\overline{W} = 1$. Figure 4.11 illustrates the evolution of a single-unit cohort, according to Equation 4.60. In the absence of fishing, the cohort peaks at $(t_p, B_p) = (5.49, .675)$. This is illustrated by the dash line. The harvesting starts earlier and proceeds relatively quickly. L is roughly 9; the whole-life harvest comes in at .6415, 95% of B_p. The Eumetric time for this μ_f is $t^* = 3.9$; the corresponding harvest h^* yields 96.1% of B_p. This is close to optimal harvesting.

Figure 4.12 illustrates the special case of yearly recruitment at the start of years $0 \rightarrow 30$, with *constant* $R = 1$. The first cohort reaches fishable maturity at $t = t_f = 3$. Subsequently, there is a buildup of biomass over a cohort lifetime, followed by a periodic steady state where annual recruitment balances annual *in situ* growth, mortality, and harvest. Recruitment ends at $t = 30$; that last cohort enters the fishable biomass at $t = 33$, and there follows a monotone decay to zero. The cumulative

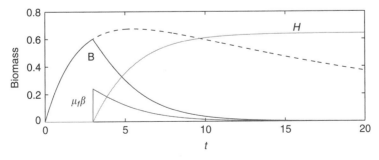

Figure 4.11. Single-cohort evolution versus time, for instantaneous unit recruitment and continuous harvesting. The dashed line is B(t) with no fishing; it is shown for reference. Also shown are the harvest rate $\mu_f\beta(t)$ and its accumulation H(t). The accumulated whole-life harvest is $H = .6415$.

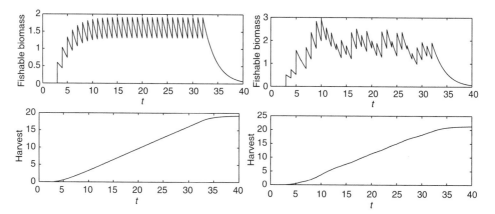

Figure 4.12. Instantaneous annual recruitment in years $0 \rightarrow 30$ and continuous harvesting: fishable biomass (*top*) and cumulative harvest (*bottom*). All cohorts are mixed and harvested concurrently with the same policy. Left: Constant $R = 1$; the annual harvest is uniform at .641 during years 15–30. As expected, this is the same as the lifetime harvest of a single-unit cohort as in Figure 4.11. Right: Random recruitment with uniform distribution $R = [0, 2]$. The actual mean $\overline{R}$ for the 30 cohorts was 1.1076; the mean annual harvest $\overline{\Delta H}$ was .7075 during the statistically steady years $15 \rightarrow 30$.

harvest asymptotes at an annual increment equal to the whole-life single-cohort harvest as illustrated in Figure 4.11.

Also shown in Figure 4.12 is the case of *random* recruitment in years 0 through 30. In this case, the uniform deviate was used for R – that is, uniform distribution of R_i between 0 and 2 with unit mean. The right panels of Figure 4.12 illustrate this case. The randomness can be clearly seen in the details, as can the structure of the response on terms of t_f, the yearly cycle, and the monotone decay following the last recruitment. A stochastic balance is achieved among recruitment, development, and harvest in roughly years 12–30. The mean annual harvest (slope of the cumulative harvest line) in this case is relatively immune to the randomness, being roughly proportional to the mean recruitment. The long fishing life of an individual cohort, $L \simeq 9$, contributes to this smoothing.

Figure 4.13 illustrates four more simulations that are statistically the same as the right half of Figure 4.12.

4.4.4   **Convolution Sum**

We can go a little further with Equation 4.70 by identifying the integral therein:

$$h_{K-i} \equiv \int_{t_K+t_f}^{t_K+t_f+\Delta t} W(t - t_i)e^{-(\mu+\mu_f)(t-t_i-t_f)}dt \tag{4.72}$$

h_j is the one-year harvest of a unit cohort ($\mathcal{R} = 1$), beginning j years after its maturation to fishable age t_f. Notice that this is strictly dependent upon the cohort vital rates and the harvesting policy (μ_f, t_f). Any biological and policy specification will define exactly $L + 1$ values of h_j, ($h_0 \rightarrow h_L$), *independent of recruitment*. With this

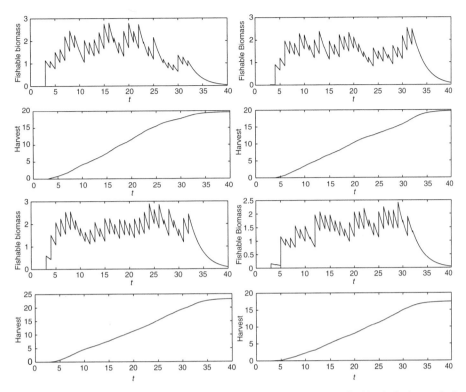

Figure 4.13. Ensemble of four simulations of stochastic cohort recruitment with identical rates as displayed in Figure 4.12. Recruitment is random, uniformly distributed on the interval $R = [0, 2]$. These outcomes are statistically identical to each other and with the stochastic outcome displayed in Figure 4.12. For the four simulations here, the values of $(\overline{R}, \overline{\Delta H})$ are (1.0215, 0.6798), (1.0273, 0.5945), (1.2077, 0.7874), and (0.9081, 0.6535).

shorthand, we arrive at the yearly harvest as convolution sum of the recruitment:

$$\Delta H(t_0) = \sum_{i=K-L}^{K} \mathcal{R}_i h_{K-i} \tag{4.73}$$

$$= \mathcal{R}_K h_0 + \mathcal{R}_{K-1} h_1 + \mathcal{R}_{K-2} h_2 + \cdots + \mathcal{R}_{K-L} h_L \tag{4.74}$$

with $t_0 \equiv t_k + t_f$.

The nature of the unit harvests h_j depends intimately on the cohort biology and the harvesting policy. Generally, h_j will ultimately become negligible as j becomes large – as we are harvesting older cohorts. Along the way, there are two competing influences in h_j: individual growth W and cohort mortality $\mu + \mu_f$. We may expect the W effect to ultimately saturate at adult weight; thereafter, there is only exponential decay and successive unit harvests h_j would decline geometrically:

$$\frac{h_{j+1}}{h_j} \simeq e^{-(\mu + \mu_f)\Delta t} \tag{4.75}$$

This relation is exact when W becomes constant in the integrand of h_j.

Steady Recruitment

It is useful here to examine the special case of *constant R*. The harvest relation (Equation 4.70) reduces in this case to

$$\Delta H(t_0) = \mu_f \mathcal{R} \sum_{i=K-L}^{K} \int_{t_0}^{t_0+\Delta t} W(t - t_i) e^{-(\mu+\mu_f)(t-t_i-t_f)} dt \qquad (4.76)$$

Each term is a one-year integral over successively aging cohorts, themselves separated in age by exactly one year and having been fished continuously since achieving age t_f. With $\mathcal{R}$ constant, this sum over identical cohorts amounts to a single continuous integral from now to eternity – the accumulated harvest over the fishable lifetime of a single cohort – H in Equation 4.9 for a single cohort. In this limit of constant $\mathcal{R}$, harvesting across cohorts is *equivalent* to lifetime harvesting of a single isolated cohort! If recruitment variability is small and its time-mean a good first approximation, then so is the single-cohort lifetime harvest, based on mean $\mathcal{R}$.

The reader is encouraged to check the equivalence of Equations 4.76 and 4.9.

In terms of the convolution coefficients, the yearly harvest with constant $\mathcal{R}$ is

$$\Delta H(t_0) = \mathcal{R} \sum h_j \qquad (4.77)$$

and it is clear that by construction, the h_j add up to the whole-life harvest of a single cohort with $\mathcal{R} = 1$.

Instantaneous Harvest

Another interesting limiting case is that of instantaneous harvest – effectively, when a cohort is completely depleted within one year of fishing, before the next cohort is recruited. This criterion might be translated into a practical limit for "high" μ_f:

$$(\mu + \mu_f)\Delta t > 4 \qquad (4.78)$$

with Δt the time between recruitment events. In this limiting case, the yearly harvest is simply $\mathcal{R}$ – essentially a lagged version of R, accounting for natural mortality and individual growth during the policy delay period t_f. There is only one term in the convolution sum. Essentially, recruitment variability is directly transferred to harvest variability. The statistical description of the time series R_i (mean, variance, autocorrelation, etc.) is the same as that for the yearly harvest, with constant amplitude scaling $e^{-\mu t_f} W(t_f)$ and pure delay t_f. Recalling Equations 4.61 and 4.62, we have three equivalent expressions:

$$\Delta H(t_0) = \mathcal{B}_K$$
$$= \mathcal{R}_K W(t_f)$$
$$= \mathcal{R}_K e^{-\mu t_f} W(t_f) \qquad (4.79)$$

with delay t_f separating cohort recruitment and harvest:

$$t_0 = t_K + t_f \qquad (4.80)$$

The three recruitment quantities in year $K - R_K$, $\mathcal{R}_K$, $\mathcal{B}_K$ are, respectively, the number of age-zero recruits, the number of those that survive to fishable age, and the biomass of those survivors.

4.4.5 Harvest Variability

For the general case of an extended harvest, we have the convolution of recruitment $\mathcal{R}$ with unit harvest h in Equation 4.74, with $L \geq 1$. Recruitment is considered an uncontrolled, random process with mean $\overline{\mathcal{R}}$, variance $\sigma_{\mathcal{R}}^2$, and zero autocorrelation. As a convolution, the annual harvest ΔH will have mean, variance, and autocorrelation, following standard relations (the overbar indicates the expected value):

$$\overline{\Delta H} = \overline{\mathcal{R}} \sum_{j=0}^{L} h_j \tag{4.81}$$

$$\sigma_{\Delta H}^2 = \overline{(\Delta H)^2} = \sigma_{\mathcal{R}}^2 \sum_{j=0}^{L} h_j^2 \tag{4.82}$$

$$^1\text{Cov}(\Delta H) = \overline{(\Delta H_t \Delta H_{t+\Delta t})} = \sigma_{\mathcal{R}}^2 \sum_{j=0}^{L-1} h_j h_{j+1} \tag{4.83}$$

The last term is the covariance of adjacent terms in the ΔH time series. It is not zero, and the autocorrelation coefficient $\rho_{\Delta H}$ is this "lag-1 covariance" normalized by $\sigma_{\Delta H}^2$. We can approximate this using the asymptote (Equation 4.75) developed above:

$$h_{j+1} h_j \simeq h_j h_j e^{-(\mu + \mu_f)\Delta t} \tag{4.84}$$

which leads to

$$\rho_{\Delta H} \simeq e^{-(\mu + \mu_f)\Delta t} \tag{4.85}$$

For the example values $\mu + \mu_f = .45$, we get the estimate $\rho_{\Delta H} \simeq .638$. The corresponding $\rho_{\mathcal{R}}$ of the recruitment time series is zero; it is by hypothesis not autocorrelated.

We can use this approximation to estimate $\sigma_{\Delta H}^2$, too. First, observe that the sum of h_j, as a unit harvest, is bounded by the weight of a fully mature individual:

$$\sum h_j \equiv \mathcal{W} < W_{\max} \tag{4.86}$$

If $h_{j+1} \simeq \rho h_j$, then[4]

$$\sum h_j \simeq h_0 \left[1 + \rho + \rho^2 + \rho^3 + \cdots \right] = h_0 \left[\frac{1}{1 - \rho} \right] = \mathcal{W} \tag{4.87}$$

$$h_0 \simeq (1 - \rho)\mathcal{W} \tag{4.88}$$

---

[4] We assume $\rho^{L+1} \simeq 0$. See Problem 10 for a more precise treatment.

Similarly, for the variance we have

$$\sum h_j^2 \simeq h_0^2 \left[1 + \rho^2 + \rho^4 + \rho^6 + \cdots \right] = h_0^2 \left[\frac{1}{1 - \rho^2} \right] \qquad (4.89)$$

The variance estimate is quickly assembled from here:

$$\sigma_{\Delta H}^2 = \sigma_{\mathcal{R}}^2 h_0^2 \left[\frac{1}{1 - \rho^2} \right] \qquad (4.90)$$

$$\simeq \sigma_{\mathcal{R}}^2 \mathcal{W}^2 \left[\frac{1 - \rho}{1 + \rho} \right] \qquad (4.91)$$

The last factor is the variance-reducing one. For the example estimate $\rho \simeq .638$, the approximate variance reduction factor $\left[\frac{1-\rho}{1+\rho} \right]$ is 0.22. Thus, 78% of the recruitment variance is eliminated in the harvest.

We can see that the mean recruitment comes through in a predictable way in the mean harvest; the sum of h_j has the sense of delivering a complete harvest, allowing for natural mortality, achieved across cohorts in any year. $\sum h_j$ is of order unity times the weight of a mature individual. The recruitment variance $\sigma_{\mathcal{R}}^2$ is muted in the harvest by the convolution as a weighted average, as expected. The longer the convolution series L, the greater is this effect. The harvest makes up for this, however, by creating covariance among *adjacent* harvests. Essentially, there is a temporal persistence in the harvest time series that is absent by hypothesis in the recruitment; it is introduced by the temporal smoothing effect of the extended harvest.

The treatment of recruitment variability and its effects on harvest is fundamental in management of wild resources and beyond this introduction. The reader is referred to several excellent sources – for example, Mangel [59].

4.4.6 Closure

Here we have developed a cohort-theoretic approach to living resource dynamics and their management. This is a special case of a stochastic system: a deterministic internal dynamic, time invariant and linear, completely uniform across all individuals and subject to uncontrolled/stochastic recruitment inputs. There are exactly two controls: the two-parameter harvest policy (μ_f, t_f). This form of harvest policy is realistic in the context of wild fishing where cohorts live longer than the recruitment interval, are mixed in nature, and cannot be separated before capture. As fixed constants to be selected, both influence harvest size, harvest variability, and the opportunity for adaptive management.

Harvest Size and Timing

Both parameters have the effect of prolonging the harvest – t_f by delaying initiation and μ_f setting the speed of the harvest once initiatied. Delay is beneficial up to the point when individual growth is offset by natural mortality in the cohort, at $t = t_p$. The largest possible harvest occurs by harvesting instantaneously at $t_f = t_p$. Any other policy falls short of this ideal harvest. There is a trade-off between t_f and μ_f: Begin earlier if μ_f is to be slower. It is never beneficial to let t_f go beyond t_p, as natural μ is an enemy. Ultimately, all cohorts will die naturally if not harvested.

Harvest Variability

Larger μ_f means a shorter fishable life L, and that means less smoothing of recruitment variability in the harvest time series. In the limit of instantaneous harvest, there is *no* smoothing: R variability is transferred directly to the harvest. Of course, this minimizes the loss to natural mortality: Biomass is harvested before it can die, but all the recruitment variability is transferred directly to the harvest. The reverse – longer L, slower μ_f – provides more R smoothing at the expense of natural mortality operative over a longer cohort life. In all cases, smaller harvest variance accompanies larger harvest persistence (autocorrelation).

Estimation and Adaptive Control

The two-parameter harvest policy is reasonable across mixed cohorts in the wild, but it *can* be adjusted over time. It is a special feature of this theory that recruitment, assumed uncontrolled, leads the harvest timewise by a pure delay t_f. It *could* be observed in the young, unfished cohorts by a suitable sampling strategy. When t_f is several years long, then there are several opportunities to estimate cohort strength while the cohort is still maturing to fishable age, even before harvesting begins. Essentially, we have the opportunity to observe R during this lag and react as best we can in programming the coming harvest.

Beyond the pure delay t_f, individual cohorts persist an additional L years, inversely proportional to $(\mu + \mu_f)$, and persist in the harvest for the full cohort life. Presumably, the harvest itself provides additional opportunities for estimating the strength of older cohorts. All of this suggests a combination of adaptive management with sequential, stochastic cohort estimation – biomass by age.

As with any "hindcast" or "nowcast" estimate, this can only improve with experience and the addition of new observations. So in that sense, long t_f provides more management opportunity. But natural mortality is always operative, so we are losing individuals to it if we wait for new data. Thus, there is a trade-off between more information and thus more policy options and more harvestable individuals. Longer harvest duration L also favors estimation during harvest; but late information has less potential for harvest policy impact.

Observation, State Estimation, Control, Forecasting, and Filtering

We have broached the notion of a stochastic system dynamics and control with potential for state estimation via observation, data assimilation and adaptive control of harvesting. There are many opportunities to add complexity and randomness – vital rates, natural variability in vital rates, harvesting policies and practices, etc.

One of the most important questions concerns the possibility of a stock recruitment relationship. If R depends even remotely on B, then short t_f and big μ_f can threaten future recruitment, as bonanza cohorts will be harvested fast, losing the staying power of a longer-lived cohort that might affect R variations. Herein we pose R to be random and uncontrolled, so there is no guidance here on this point.

4.5   A COHORT OF INDIVIDUALS

We began this chapter with the goal of studying cohorts of "identical individuals." Considerable insight has been gained from separate descriptions of individual vital processes affecting their maturity W and mortality affecting their number N.

But all individuals are not the same. Variability is observed across individuals in essentially all vital processes. This obvious fact deserves articulation.

4.5.1   Individual-Based Processes

Growth Rate Distribution

As a first step, one can define discrete cohorts of *statistically* identical individuals, but with variability in vital rates specified as distributions – probability density functions with mean, variance, skewness, etc.

For example, consider the growth function introduced at Equation 4.16 and used heavily herein:

$$G(W) = g\left(W^* - W\right) \tag{4.92}$$

(Here we have renamed the limiting value W^* and reserve overlining to indicate an average.) Integration for a single individual, starting at $W = 0$, gives the weight function

$$W(t) = W^*\left[1 - e^{-gt}\right] \tag{4.93}$$

There are two parameters, g and W^*. Both can be expected to vary among otherwise identical individuals.

Suppose the constant g varies continuously across individuals, with the *uniform* distribuion over the range $g_i = g_0 \pm \Delta g$. Then we have a range of slow- and fast-growing individuals, each with identical mature weight W^*. The mean over all individuals at a fixed time will be

$$\overline{W}(t) = \frac{\int W(t)dg}{\int dg} \tag{4.94}$$

with an integration range from $(g_0 - \Delta g)$ to $(g_0 + \Delta g)$. The result:

$$\overline{W}(t) = W^*\left[1 - e^{-gt}\left(\frac{\sinh(t\Delta g)}{t\Delta g}\right)\right] \tag{4.95}$$

(Sinh is the hyperbolic sine function; $\sinh(x)/x \geq 1$ assuming $x \geq 0$.) The cohort, on average, will mature more slowly than a single individual with the average g. Growth essentially ceases as W^* is reached; the average individual reaches maturity while the slower individuals are still growing.

Program **IBMa.xls** illustrates this. The reader is encouraged to verify the result, Equation 4.95, in Problem 16.

This simple example opens many complexities: the need for a fuller description of g's variability in terms of its probability density function; the similar need for a statistical description of W^* variability; how g and W^* covary; their variabilities over time; their autocorrelation; etc. While fruitful, this rapidly gets out of hand with respect to analysis alone. One seeks an approach to variability that is simulation oriented: Put the variability into the parameters, across individuals; simulate each individual; then aggregate the *results* as needed. Such an approach would keep the full variability of the ensemble present across all individuals.

$$\frac{dW_i}{dt} = G_i(W_i) \tag{4.96}$$

This individual-based approach is feasible on contemporary machines. Essentially: Simulate vital processes of many diverse individuals, separately; and sum the results over individuals to find their aggregate properties.

Discrete Mortality Events

It is useful to conceive of certain individual processes as having discrete, binary outcomes. Mortality is a good example. If the probability of death in an interval Δt is $\mu \Delta t$, then the probability of survival P is its complement:

$$P(\Delta t) = 1 - \mu \Delta t \tag{4.97}$$

and survival to time $t = k\Delta t$ has probability

$$P(k\Delta t) = [1 - \mu \Delta t]^k \tag{4.98}$$

and with $k\Delta t \equiv t$,

$$P(t) = \left[1 - \frac{\mu t}{k}\right]^k \tag{4.99}$$

Keeping $t = k\Delta t$ constant but increasing the resolution such that $\Delta t \to 0$, we have the limit

$$P(t) = \lim_{k \to \infty} \left[1 - \frac{\mu t}{k}\right]^k = e^{-\mu t} \tag{4.100}$$

At high resolution, we find exponential survivorship with decay rate μ. Discrete survivorship converges to the continuous form

$$\frac{dP}{dt} = -\mu P \tag{4.101}$$

Suppose mortality were acting simultaneously on a cohort of N individuals. Then the survival probality of each would aggregate to the number of surviving individuals:

$$N(t) = N_0 \left[1 - \frac{\mu t}{k} \right]^k \tag{4.102}$$

or its continuous limit. This is the essence of individual-based representation of mortality: Assign a binary probability of survival as in Equation 4.97 for the small duraton Δt; allow individual survival consistent with that probability, for all individuals separately. The aggregate effect on the population will achieve the desired mortality.

The uniform deviate[5] $\mathcal{U}$ may be invoked to do the binary sorting. Suppose an individual has probality $\mu \Delta t$ of death. Then a random deviate from the uniform distribution can be used to assign survival for that individual. For example, if

$$\mathcal{U} > \mu \Delta t \tag{4.103}$$

then the individual survives; otherwise, not.

This places a resolution requirement on the selection of Δt – clearly, $\mu \Delta t$ must be kept small (and certainly below 1) or results will be sensitive to Δt.

Residence Time and Stage Transition

Suppose there are N individuals in a box and that they escape at the rate αN:

$$\frac{dN}{dt} = -\alpha N \tag{4.104}$$

with solution

$$N(t) = N_0 e^{-\alpha t} \tag{4.105}$$

Many processes are represented in this way (e.g., mortality, as above). A common application relates to individual *stage duration* for populations that progress through a series of stages.

At any point in time t, there are $\alpha N(t) \Delta t$ individuals escaping, and each has "age" t. The average age of escapees over the interval $0 \leq t \leq \tau$ is

$$\overline{T} = \frac{1}{N_0 - N(\tau)} \int_{t=0}^{t=\tau} t \cdot \alpha N dt \tag{4.106}$$

We have $N(t)$ above; its substitution allows integration of Equation 4.106. In the limit as $\tau \to \infty$, all individuals escape, and their average residence time is

$$\overline{T} \to \frac{1}{\alpha} \tag{4.107}$$

---

[5]   See Appendix for a brief treatment of random numbers.

Other moments can be similarly evaluated. The complete distribution of time-in-stage turns out to be the exponential, or waiting time, distribution, discussed in the Appendix. This distribution is skewed; there are no negative values; small values are the most common, but very high values are possible with diminishing probability. Acccordingly, the median is lower than the mean.

The summary result: A cohort with first-order residence in stage as in Equation 4.104, has residence times distributed according to the exponential distribution with parameter α. The properties are (a) **mean** $\overline{T} \equiv 1/\alpha$; (b) **variance** = $\overline{T}^2$; and (c) **median** = $.69\overline{T}$. It would be common to observe the outcome, estimate $\overline{T}$, and represent the process as a distribution of individual waiting times.

From an individual's point of view, the discrete probability P of exit from stage during the next Δt is

$$P \approx \alpha \Delta t = \frac{1}{\overline{T}} \Delta t \qquad (4.108)$$

The approximation improves as $\Delta t \to 0$. This probability of escape is independent of prior events. Specifically, it does not depend on the time already spent in residence. This is a remarkable feature of this distribution. It facilitates implementation of the waiting time distribution on a very local basis: at the level of the individual undergoing a single small time step Δt. The aggregate properties result.

So variability in residence time can be realized at the individual level simply by assigning the probability of exiting the stage and making a binary statistical determination for each individual (as in Equation 4.103). The cohort will have the proper aggregate properties, with the desired distribution fully represented at the individual level.

Operationally, stage completion for an individual is the same as individual mortality. They both invoke the same discrete simulation apparatus even though the two processes correspond to very different realities.

Reproduction

Reproduction is another binary process when considered from the individual's view. Suppose the aggregate reproduction rate is

$$\frac{dN_0}{dt} = \rho N_a \qquad (4.109)$$

with recruitment into stage 0, and N_0 the number of individuals in that earliest life stage, and N_a the number of reproductive-stage individuals. For an individual within the N_a cohort, the probability of reproduction in the period Δt would be

$$P = \rho \Delta t \qquad (4.110)$$

Given the probability, reproduction is a binary matter: to generate either 1 or 0 new recruits for a given adult in a given Δt. The summation over all reproductive-stage individuals would lead to the correct number of recruits $\rho N_a \Delta t$. The same binary apparatus invoked in the other processes is relevant here. Initializing a new individual can be done in a way that inherits properties from the individual parent.

For populations with sexual reproduction, male and female individuals would be differentiated randomly at an appropriate age, with reproductive capability assigned only to female individuals. It is common to ignore the problem of sexual contact among individuals, assuming it does not limit the process.

Motion

Motion of concentrations or densities of particles is commonly represented by the sum of advection and nonadvective portions. The advective part is common to all individuals at a given location; the nonadvective part is typically represented as a diffusive process, itself the aggregate of many random motions with zero mean. Each individual will experience a different random sample of the diffusive part.

Individual motion is therefore represented as the sum of advection and a random walk:

$$X_i^{k+1} = X_i^k + V(X_i) \cdot \Delta t + \epsilon \sqrt{2D\Delta t} \qquad (4.111)$$

with ϵ a zero-mean, unit-variance random variable and V the deterministic advective velocity. The random walk is configured so that the aggregate over many individuals is equivalent to diffusion; D is the diffusion coefficient. There are additional terms required when D varies with X. (See, for example, [73].)

These motions are physical in origin; there is additional motion Vb due to behavior. Examples include light- or oxygen-seeking, predator avoidance, diel migration, etc. Vb will develop as does the organism; it will certainly be subject to physical constraints (effects of heat or turbulence) as well as biological ones – for example, constraints on speed. Overall, the individual motion would be governed by a common Lagrangian integral with random walk:

$$X_i^{k+1} = X_i^k + [V(X_i) + Vb_i] \Delta t + \epsilon \sqrt{2D\Delta t} \qquad (4.112)$$

One requires a single random deviate ϵ, plus externally specified physical fields (V, D) plus biological formulation for Vb in terms of the organism's current state of development.

4.5.2 Individual-Based Simulation

In simulation, one breaks all processes into the simplest describable units. In the present case, this amounts to describing a single individual evolving over one time step. The aggregation takes over from there, generating a cohort with individual outcomes reflective of its initial variability.

Some cohort processes are really modeled as individual processes – for example, individual weight and individual location. These retain their character as continuous variables requiring integration over time. Individuals vary due to parameter variation; and, as in the case of random walk, by stochastic forcing. But the canonical problem is the integration of conventional differential equations for W or X over a single time step.

There are other processes that are binary at the individual level. Included are mortality, stage transition, and reproduction. In these cases, the procedure for an individual would be to (a) form the probability of an event during the coming Δt; (b) generate the uniform deviate $\mathcal{U}$; and (c) accept or reject the event based on the comparison.

When the vital rate parameters – for example, growth rates, mortalities, reproduction rates – depend on X, then we have "physical-biological" coupling between motion and growth.

A simple example cohort might have the following state variables:

- Real variables
  - weight
  - "condition" or health (possibly a convolution of history)
  - time since entering stage
  - position (a pointer to the environment)
- Integer variables
  - alive or dead (binary)
  - male or female (binary)
  - stage
  - number of prior reproductive events

4.5.3 Spatially Explicit Populations

Physical-biological coupling is most obvious in spatially explicit simulations, where a realistic model environment – heat, light, nutrients, predators, motion – is inhabited by a model cohort.

In many practical cases, the knowledge of physics supports detailed site-specific simulations that are not affected by the biology, while the field scale biology is still in the hypothesis-testing phase. This is most apparent when one is concerned with manipulation of harvesting regimes or protecting habitat at the field scale. There is much value in detailed, spatially explicit simulation of model cohorts in realistic environments in this case. Recent marine examples using a common physical model include populations of scallops, zooplankton, larval fish, lobsters, and phytoplankton (see below). These simulations have the possibility of achieving field scale population abundance and distribution, constrained by annual variations in current, temperature, wind, etc. and their effects on aggregation, dispersal, and vital rates.

There is a great deal to be said about individual-based modeling. Here we call attention to the potential of this general approach. It embeds elements of both cohort models and stage-structured models of populations; and it allows coupling to potentially realistic environments. The resulting coupling between physics and biology is critical when the environment varies spatially and temporally on scales comparable to that observed in the population itself.

4.6 RECAP

The cohort analysis presented here uses a single, continuous life stage. It would be natural to generalize this to the case of multiple stages, with weight gain in each stage. The population literature cited in the previous chapter generally explores the union of cohort- and stage-based analyses.

The natural extension is to individual-based representation, in order to recognize diversity in vital rates and behavior within cohorts. This is introduced here and is possible on today's machines. Early general work is represented in Sammarco and Heron [78], Grimm and Railsback [36], and Deangelis and Gross [17]. Werner et al. [94] provide a useful review of recent research. The important influence of diffusion in realistic environments is explored by Okubo and Levin [69]. Clark and Mangel [11] provide a detailed look at behavioral development (a rework of the earlier work [58]).

Real management options are site specific – that is, they address specific populations in specific physical environments. The individual-based model can be a powerful tool in bringing theory and observation together to bear on real problems. A site-specific case study, with a single physical model and several different populations, can be followed in the work of Proehl et al. [73] (dispersion), Miller et al. [64] (zooplankton), Tremblay et al. [89] (shellfish), Werner, Lough et al. [93, 92, 55, 54] (cod), and McGillicuddy et al. [62] (phytoplankton).

4.7 PROGRAMS

- **Cohort1.xls** simulates the vital rates for the exponential example.
- **Cohort2.xls** adds the Eumetric yield curve.
- **Forest3-05.xls** evaluates the Forester and Faustman rotations versus rotation period. The data are from Neher [68], Chapter 2, for Douglas fir.
- **CoMass1.m** simulates an isolated cohort, as in the example for uncontrolled recruitment.
- **CoMass2.m** simulates multiple mixed cohorts, recruited annually but with random size. This is done by convolution of single cohorts of random size.
- **IBMa.xls** evaluates the effect of parameter variability on ensemble average (Equation 4.95).

4.8 PROBLEMS

1 Given individual growth according to

$$\frac{dW}{dt} = G(W) = g\overline{W}\left(1 - W/\overline{W}\right) \tag{4.113}$$

with G the growth rate, $W(t)$ the individual weight, and parameters $(g, \overline{W})$ as in Section 4.2. Show that the weight $W(t)$ is

$$W(t) = \overline{W}\left(1 - e^{-gt}\right) + W_0 e^{-gt} \tag{4.114}$$

where W_0 is the initial weight at $t = 0$.

2 Continuation of Problem 1. Find W_p, t_p, B_p, and h_p for $W_0 \neq 0$. Compare with the text results for $W_0 = 0$ (Equations 4.17 ff).

3 Suppose individual growth is governed by the logistic equation

$$\frac{dW}{dt} = G(W) = gW\left(1 - W/K\right) \tag{4.115}$$

with constant parameters K and g. Confirm that the solution is

$$W(t) = \frac{K}{1 + C_0 e^{-gt}} \tag{4.116}$$

with the constant C_0 given from the initial condition W_0:

$$C_0 = \frac{K - W_0}{W_0} \tag{4.117}$$

4 For logistic growth, as in Problem 3, and natural mortality given by

$$\frac{dN}{dt} = -\mu N \tag{4.118}$$

(a) Find t_p (the time at which the biomass peaks), the individual weight W at t_p, and the peak biomass B_p.

(b) What happens if $\mu > g$?

5 Continuation of Problem 4:

(a) Plot $W(t)$, $N(t)$, and $B(t)$, as in Figure 4.1, for $[g, R, K, W_0] = [.01, 1, 1, .05]$. Assume no harvesting.

(b) Confirm graphically your answers to Problem 4.

6 A cohort undergoes logistic individual growth, with natural mortality rate μ. Show that the number of survivors when the biomass peaks is

$$N(t_p) = R\left[\frac{\mu}{C_0(g - \mu)}\right]^{\mu/g} \tag{4.119}$$

where C_0 is determined from initial weight W_0, as in Problem 3:

$$C_0 = \frac{K - W_0}{W_0} \tag{4.120}$$

7 (Solve Problem 4 first.) Recruits R are purchased at weight W_0, and grow at the rate

$$G = g\overline{W}\left(1 - W/\overline{W}\right) \tag{4.121}$$

The cost of each recruit C depends on its weight:

$$C(W_0) = C_0 W_0 + C_1 \tag{4.122}$$

There is no mortality. Harvesting is instantaneous as B peaks and is sold at value $P \cdot B$. There is no subsequent recruitment. Interest rate is r.
   (a) Express the present worth of this activity, as a function of W_0.
   (b) What value of W_0 maximizes the present worth?
   (c) Plot the result of (a). Does the plot agree with your answer to (b)?

8 Consider the cohort biomass function

$$B(t) = t^2\left(1 - \frac{t}{K}\right) \tag{4.123}$$

with t the cohort age in years and $K = 100$ years. For costless harvesting,
   (a) Evaluate the Forester's ideal harvest age.
   (b) Plot the Forester's sustained yield (biomass per year) as a function of harvest age. Do you get a maximum and does it agree with the above?
   (c) Evaluate the Faustman ideal harvest age as a function of interest rate r.
   (d) Plot the Faustman rent as a function of harvest age, for $r = 0.025, 0.05, 0.075, 0.10,$ and 0.15 per year. Do the peaks agree with the Faustman optima?
   (e) Plot versus interest rate the Faustman harvest time, the Faustman rent, and the Faustman annuity.

9 Derive Equations 4.20, 4.23, 4.24, and 4.25.

10 In Equations 4.87 and 4.89, we used the approximation

$$\sum_{j=0}^{L} \kappa^j \simeq \frac{1}{1-\kappa} \tag{4.124}$$

for $\kappa^2 < 1$. This is valid for large L and leads to the simple expression in Equation 4.91. But more precisely,

$$\sum_{j=0}^{L} \kappa^j = \frac{1 - \kappa^{L+1}}{1-\kappa} \tag{4.125}$$

Using this, rework Equation 4.91.

11 A fishery with set harvesting policy is characterized by the following one-year harvests for a unit cohort: $(h_0, h_1, h_2, h_3, h_4, h_5, h_6) = (8, 10, 8, 4, 2, 1, 0.5)$. Recruitment has

mean 100, standard deviation 60, and zero autocorrelation. Natural mortality is $\mu = 5\%$. Estimate
- (a) the fishing mortality (μ_f)
- (b) the weight of a mature fish $W_{\max}$
- (c) the mean annual harvest $\overline{\Delta H}$
- (d) the standard deviation of the annual harvest $\sigma_{\Delta H}$
- (e) the autocorrelation of the annual harvest $\rho_{\Delta H}$

12 A forest is characterized by

$$B(t) = B_0 \sqrt{t - 20} \qquad (4.126)$$

where $B(t)$ is the mass of marketable timber for a cohort of age t. For $t < 20$, $B = 0$,
- (a) Plot $B(t)$.
- (b) Using the Forester criterion, what is the optimal rotation period? the optimal annual supply of timber mass?
- (c) Repeat (b) using the Faustman rotation. Plot the answers versus interest rate r, for several values: .025, .05, .10, .15, and .20 per year.

13 The harvestable volume of timber (B, board feet) in a private forest is given by

$$B(t) = 1{,}200t^2 \left(1 - \frac{t}{100}\right) \qquad (4.127)$$

with t the forest age (years) since planting. Harvesting is costless; the interest rate r is 5% per year.
- (a) Plot B versus t.
- (b) At what age does the forest reach maximum harvestable volume? What is that volume?
- (c) What is the age for maximum harvest per year? If this is different from the answer to (b), explain why. What volume is harvested at that point?
- (d) What are the Faustman rotation period and the volume harvested? Explain the differences among answers to (b), (c), and (d).

14 A forest grows according to

$$B(t) = t^2 e^{-at} \qquad (4.128)$$

with $B(t)$ the accumulated board feet at age t and $a = 3\%$ per year. Harvesting is costless; the value of the timber is constant at \$5 per board foot. The interest rate is 5% per year.
- (a) For a single harvest: What is the maximum yield in board feet? When does it occur?
- (b) What is the maximum Sustainable yield (Forester's rotation) in board feet per year? What is the rotation period?
- (c) For a single harvest: What age at harvest will maximize the present worth? What is the present worth of that harvest?
- (d) For a sustainable sequence of harvests: Repeat (c).

(e) Reconsidering (d): Suppose there is a large number of identical forests under identical management, except that they are started at random times. What is the sustainable yield in board feet per year?

(f) The interest rate just went up to 15% per year. Redo all the above (a–e) and explain.

15 Individuals are bred in a nursery and recruited at weight W_0. Thereafter, they undergo logistic growth and experience natural mortality.

(a) At what age and weight would harvesting occur when using the Forester criterion? Express your answers as functions of W_0 and in terms of the parameters K, g, and μ.

(b) The same, but for the Faustman criterion.

16 Derive Equation 4.95.

17 Compare continuous and discrete mortality processes for $\mu = 0.1$, $\Delta t = 1$.

(a) Compute and plot discrete and continuous probabilities of survival as a function of time for $0 \leq t \leq 40$.

(b) Compute and plot the difference between them as a function of time.

(c) Repeat (a) and (b) for $\mu = 0.2$ and for $\mu = 0.3$.

(d) Do the two different representations agree when $\mu \Delta t$ is small?

18 Starting from Equation 4.105: Derive the result (Equation 4.107) for the mean waiting time for a first-order decay process. Fill in the steps not recorded in the text.

19 Generate waiting times with the exponential distribution:

(a) Show that the generator $T = -\overline{T} \ln(\mathcal{U})$, with $\mathcal{U}$ the uniform deviate, produces the exponential distribution for waiting time T with mean $\overline{T}$. What ensemble size is needed to reproduce this property?

(b) Show that the variance of T is $\overline{T}^2$. What ensemble size is needed to reproduce this property?

(c) Plot mean, standard deviation, and median as functions of ensemble size.

20 A life stage has an absolute minimum duration of three years; beyond that, the median stage duration is one additional year, and the waiting time (exponential) distribution is valid. There is no mortality.

(a) What is the mean overall stage duration? What is the median?

(b) Construct a discrete model for individual stage duration.

(c) A group of 10 individuals enters the stage at $t = 0$. Simulate this stage; generate a completion time for each individual. Are the results in accord with theory?

(d) Repeat (c) N times. Compute the mean, median, and standard deviation of the stage durations as a function of N. Do you recover the result expected, and how big need N be to reliably estimate these properties?

21 Repeat Problem 20(b) through (d), with mortality as follows: $\mu = [.1, .05, 0]$ in years 1, 2, and 3; and thereafter, $\mu = .05$.

5 Water

Water is representative of Quadrant 2: the sterile, renewable resource. It is fundamentally fugitive, although it moves in predictable patterns. It is renewed by the planetary evaporation–precipitation process. One is concerned with its rate of occurrence Q, which occurs in a terrestrial network. Consumptive use evaporates water and returns it to the atmosphere; nonconsumptive use creates no vapor, but returns used water to the network for reuse. One thus distinguishes consumption and withdrawal. In important ways, these differ; both exhibit "networked rivalry" in different ways.

There are both steady and transient analyses that are important. Both use the same network, and both require capital investment. In the steady case, variability in either rainfall or upstream use is passed directly to downstream riparians. In the transient case, one adds intentional modification of this variability in order to manipulate or react to variability. Both cases require important resource management activities; both imply the deployment of significant capital resources in order to maintain and improve the natural distribution system. And all uses require a constant vigilance to ensure that nonconsumptive use, when returned, is not degraded in quality.

5.1 INTRODUCTION

How do we conceive the water resource? It is distributed, constantly renewed by a planetary geophysical distillery, invoking ocean, atmosphere, and sun. It is deposited as precipitation in catchments in which we live, defined by continental relief and orientation.

There is no question about the scarcity of water. Chemically, man is more that half H_2O and the residence time of an individual molecule within an individual human is about a month. Access to a constantly renewed flow of fresh water is essential to human life. In occurrence, the water resource is renewable, sterile, distributed, massive, and specific to location and time. It is fugitive, constantly in motion. It is incapable of being isolated in any meaningful quantity for very long; but it is expendable (consumable) by conversion back to vapor, potentially anywhere. Similarly, it is degradable by pollution. Both "none" and "plenty" are characteristics of its stochastic occurrence: It can be too abundant for habitation; its absence makes life impossible. Overall, people conform their activities to water's occurrence and attempt to mitigate its idiosyncrasies.

Human settlements rely on the natural distribution of water, with some modifications. Care in its geographic *transfer* is always called for: It is energetic and constitutes the habitat for plants and animals as well as humans. It colors human culture in its natural manifestations (for example, lakes, rivers, and deserts). Its removal can lead to a life-threatening shortage. It is best to think modestly about how to utilize given natural systems of delivery in contemplating the human prospect.

Overall, the basic rule of thumb is that water occurrence is renewable and the human need is persistent. So, flowrates matter, not quantities stored. Water is sterile (but degradable by pollution), and its flows occur in patterns that are highly conditioned by geographic features. Civilizations form around it and are constrained by it; development is guided by its necessity and potential.

5.2 WATER AS A PRODUCTIVE RESOURCE

To begin with, consider a simple farm system in which water is essential to crop growth. Its supply is limited (W), and there are two other scarce resources, land (L) and fertilizer (F). We imagine growing two crops, cotton and barley, in to-be-determined quantities (X_c, X_b). Each has a net value (V_c, V_b) per ton. The per-ton amounts of resources (water, land, fertilizer) needed for cotton are known: (10, 1, 2). Similarly, for wheat the requirements per ton are (5, 1.5, 2). By inspection, cotton is water intensive (10:1) relative to barley (5:1.5), which is relatively land intensive. These data are summarized in Table 5.1.

The constraint of the problem is simply to avoid crop mixes (X_c, X_w) that require more resources than available:

$$10X_c + 5X_b \leq W \tag{5.1}$$

$$1X_c + 1.5X_b \leq L \tag{5.2}$$

$$2X_c + 2X_b \leq F \tag{5.3}$$

Table 5.1. Farm data with two crops and three resources.

	Cotton	Barley	Resource Limit
Water	10	5	W
Land	1	1.5	L
Fertilizer	2	2	F
Crop Amount	X_c	X_b	≥ 0
Crop Value	V_c	V_b	–

There are no other limits on the crop amounts, but they must not be negative:

$$X_c \geq 0 \tag{5.4}$$

$$X_b \geq 0 \tag{5.5}$$

and the value Z of the activity is

$$V_c X_c + V_b X_b = Z \tag{5.6}$$

The production planning problem we envision: Maximize crop value Z within the limits of the resource constraints in Equations 5.1–5.3 and subject to the nonmeaning of negative cropping (Equations 5.4 and 5.5).[1] This is well posed, provided the five data for resouce availability (W, L, F) and crop value (V_c, V_b) are known.

5.2.1 Water and Land

Figure 5.1 illustrates the crop production possibilities (X_c, X_b) for the simple case of unlimited fertilizer. Two constraints are important and potentially binding: water

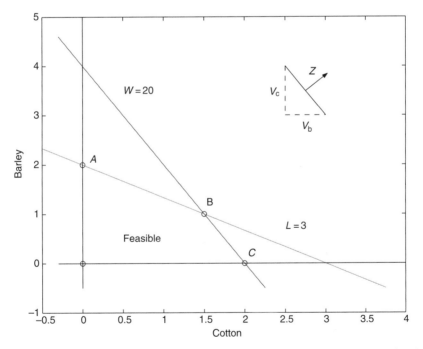

Figure 5.1. Production space (X_c, X_b) and constraints on water ($W = 20$) and land ($L = 3$); fertilizer is unlimited; production is limited to positive values. A contour of constant value Z is shown, a straight line; and the direction of increasing Z is indicated. The circles indicate the four possible extrema for a linear objective Z. The region of feasible solutions is labeled; it is enclosed by the convex quadrilateral with the circled corners.

---

[1] If this farm were allowed to import crops, then negative X would become meaningful.

and land. There is a quadrilateral enclosing the feasible solutions, defined by the four vertices marked. Each of these is the intersection of exactly two of the constraints.

As illustrated, value Z increases toward the northeast, assuming positive values for (V_c, V_b); their ratio determines the slope of the isolines (contours). Clearly, the origin is the do-nothing alternative; and value increases as one moves outward into the first quadrant and across the feasible space. The outward motion ultimately is halted at the boundary of feasibility. Because Z is linear in the decisions (X_c, X_b), the maximum of Z will occur at one of the vertices marked. *At each such vertex, exactly two constraints will be binding*, (the equation is a strict equality), and the other constraints will be slack (the equation is a strict *inequality*). Excluding the origin, there are three vertices, and they are distinguished by (a) what the crop selection is; (b) what constraints bind; and (c) how much resource is left unused, or "slack."

For each of the three points, there is a range of relative value V_c/V_b that favors it. Clearly, very high V_c will select point C (all cotton); very low V_c, point A (all barley); and intermediate V_c, point B (a blend). These ranges are shown in Table 5.2.

Because there are only three interesting solutions, we can go further. Visually, each is the intersection of exactly two binding constraints, with the others left slack – that is, satisfied but not binding as equalities. This is a useful way of characterizing the three points. For example, at point B, the binding constraints are water and land:

$$\begin{bmatrix} 10 & 5 \\ 1 & 1.5 \end{bmatrix} \begin{Bmatrix} X_c \\ X_b \end{Bmatrix} = \begin{Bmatrix} 20 \\ 3 \end{Bmatrix} \tag{5.7}$$

and we quickly confirm that the solution is

$$\begin{Bmatrix} X_c \\ X_b \end{Bmatrix} = \frac{1}{10} \begin{bmatrix} 1.5 & -5 \\ -1 & 10 \end{bmatrix} \begin{Bmatrix} 20 \\ 3 \end{Bmatrix} = \begin{Bmatrix} 1.5 \\ 1 \end{Bmatrix} \tag{5.8}$$

or equivalently,

$$\begin{Bmatrix} X_c \\ X_b \end{Bmatrix} = \begin{bmatrix} .15 & -0.5 \\ -.10 & 1.0 \end{bmatrix} \begin{Bmatrix} W \\ L \end{Bmatrix} \tag{5.9}$$

Table 5.2. Farm solutions with two crops and three resources. In this example, fertilizer is unlimited, but $W = 20$ and $L = 3$. The three solutions, A, B, and C, are marked on Figure 5.1.

	A		B		C	
Program	Amount	Slack	Amount	Slack	Amount	Slack
Cotton	0	–	1.5	–	2	–
Barley	2	–	1	–	0	–
Water	10	10	20	0	20	0
Land	3	0	3	0	2	1
Fertilizer	4	∞	5	∞	4	∞
V_c/V_b		<2/3		[2/3 – 2]		>2

Locally, the values W and L control the solution. If they were to be altered by the small amounts $(\Delta W, \Delta L)$, the impact on the solution would be

$$\begin{Bmatrix} \Delta X_c \\ \Delta X_b \end{Bmatrix} = \begin{bmatrix} .15 & -0.5 \\ -.10 & 1.0 \end{bmatrix} \begin{Bmatrix} \Delta W \\ \Delta L \end{Bmatrix} \tag{5.10}$$

$$= \begin{Bmatrix} .15 \\ -.10 \end{Bmatrix} \Delta W + \begin{Bmatrix} -0.5 \\ 1.0 \end{Bmatrix} \Delta L \tag{5.11}$$

Immediately, the trade-off between the two crops is apparent, if either constraining resource were to be altered. If ΔW is positive, the production change is along the land constraint; this is apparent in Figure 5.1 by a visual perturbation. Similarly, an alteration ΔL causes the crop mix to move along the water constraint.

Changing a resource constraint alters production and hence value. At the marginal values (V_c, V_b), we have an overall impact

$$\Delta Z = V_c \Delta X_c + V_b \Delta X_b \tag{5.12}$$

$$\Delta Z = [.15V_c - .10V_b]\Delta W + [-.5V_c + V_b]\Delta L \tag{5.13}$$

The quantity $[.15V_c - .10V_b]$ is the *shadow price* of water in this instance – the inpact on value of a unit change in water availability. The range of relative value V_c/V_b (Table 5.1) that makes vertex B optimal, guarantees that this shadow price is nonnegative. Similarly, the quatity $[-.5V_c + V_b]$ is the nonnegative shadow price of land. Generally, every binding constraint will have an associated nonzero shadow price. Nonbinding constraints are exactly that – they do not affect the solution locally; hence, their shadow prices are formally zero.

This calculation is specific to the vertex in question. At point A, for example, the two binding constraints are land and the nonnegativity of the cotton crop. The matrix there is

$$\begin{bmatrix} 1 & 1.5 \\ 1 & 0 \end{bmatrix} \begin{Bmatrix} X_c \\ X_b \end{Bmatrix} = \begin{Bmatrix} L \\ 0 \end{Bmatrix} \tag{5.14}$$

and we quickly confirm that the sensitivity to ΔL is

$$\begin{Bmatrix} X_c \\ X_b \end{Bmatrix} = \begin{Bmatrix} 0 \\ \frac{2}{3} \end{Bmatrix} L \tag{5.15}$$

and

$$\begin{Bmatrix} \Delta X_c \\ \Delta X_b \end{Bmatrix} = \begin{Bmatrix} 0 \\ \frac{2}{3} \end{Bmatrix} \Delta L \tag{5.16}$$

$$\Delta Z = \left[\frac{2}{3} V_b \right] \Delta L \tag{5.17}$$

The shadow price of land is $\left[\frac{2}{3} V_b \right]$. Water is slack at point A; there is a surplus, and therefore its shadow price is null.

Table 5.3. The same three solutions A, B, C as in Table 5.2, from the resource viewpoint. P is the shadow price of the resource. The three solutions A, B, C are as marked on Figure 5.1.

Program	A			B			C		
	Used	Slack	P	Used	Slack	P	Used	Slack	P
Water	10	10	0	20	0	$[.15V_c - .10V_b]$	20	0	$.10V_c$
Land	3	0	$.67V_b$	3	0	$[-.50V_c + V_b]$	2	1	0
Fertilizer	4	∞	0	5	∞	0	4	∞	0
V_c/V_b		<2/3			$[2/3 - 2]$			>2	

At point C, the reverse is true: Water constrains; there is a surplus of land; and barley is driven to zero. It is easy to confirm that at point C the governing equations are

$$\begin{bmatrix} 10 & 5 \\ 0 & 1 \end{bmatrix} \begin{Bmatrix} X_c \\ X_b \end{Bmatrix} = \begin{Bmatrix} W \\ 0 \end{Bmatrix} \tag{5.18}$$

The solution:

$$\begin{Bmatrix} X_c \\ X_b \end{Bmatrix} = \begin{Bmatrix} 0.1 \\ 0 \end{Bmatrix} W \tag{5.19}$$

$$\begin{Bmatrix} \Delta X_c \\ \Delta X_b \end{Bmatrix} = \begin{Bmatrix} 0.1 \\ 0 \end{Bmatrix} \Delta W \tag{5.20}$$

$$\Delta Z = [0.1V_c]\Delta W \tag{5.21}$$

The shadow price of water is $[0.1V_c]$; for land, there is a surplus, and its shadow price is null.

Table 5.3 summarizes the three solutions A, B, C from the *resource* point of view – the amount of the resource used, the amount left over (slack, S) and the marginal value or shadow price P of the resource. The binding constraints have zero slack and nonzero shadow price. The reverse is true of the nonbinding constraints. This "dual" view of the system illustrates the "complementary slackness principle": For all resources, $S \cdot P = 0$.

5.2.2 Adding a Resource

The preceding solutions were reached with unlimited fertilizer supply. Clearly, if F is finite, then it can also exert influence on the solutions. Point B above consumed 5 units of fertilizer. If the limit $F < 5$, then at least this solution is not feasible; and we need to consider the F limit synergistically with water and land. In Figure 5.2, we have added the limit for $F = 4.5$ to the picture. Clearly, point B is no longer feasible, and two new adjacent vertices B1 and B2 have become important. The set of interesting solutions for this case is summarized in Table 5.4.

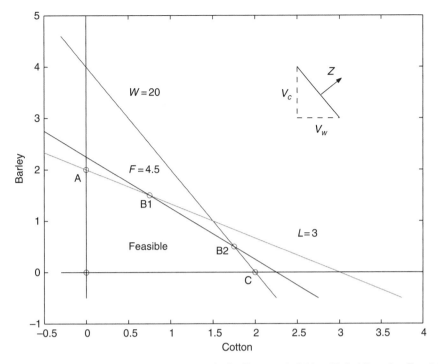

Figure 5.2. The Same as Figure 5.1, except the fertilizer constraint is added at the value $F = 4.5$, further restricting the feasible space. There are now five vertices defining the convex feasible polygon, marked by the circles.

Table 5.4. Farm data with two crops and three resources. Fertilizer is limited, $F = 4.5$; and, as above, $W = 20$, $L = 3$. Points A and C are the same.

	A		B1		B2		C	
Program	Amount	Slack	Amount	Slack	Amount	Slack	Amount	Slack
Cotton	0	–	.75	–	1.75	–	2	–
Barley	2	–	1.5	–	.5	–	0	–
Water	10	10	15	5	20	0	20	0
Land	3	0	3	0	2.5	.5	2	1
Fertilizer	4	.5	4.5	0	4.5	0	4	.5
V_c/V_b	<2/3		[2/3 – 1]		[1 – 2]		>2	

Points A and C remain unchanged. The new point B1 is characterized by the binding of constraints on fertilizer and land; those resources will have nonzero shadow prices. Analogously, B2 has binding constraints on fertilizer and water. The reader is encouraged to develop the full solutions at these points and to confirm the range of relative values V_c/V_b that applies to each.

This is an instance of the canonical linear-programming problem, where all constraints and the objective are linear and the decision variables are real. There are many

implementations; one useful student package is bundled within Excel, utilizing the
Solver tool. The program **WaterLand.xls** is an implementation of the specific problem described here. The student is encouraged to explore all the features developed
above.

5.2.3 Canonical Forms

Clearly, the above formulation can be extended to include more resources and/or
more crops or other resource-consuming activities. The general linear form[2] is

$$[A]\{X\} \le \{R\} \tag{5.22}$$

$$\{X\} \ge \{0\} \tag{5.23}$$

$$\{V\}^T\{X\} = Z \tag{5.24}$$

for the constraints; the objective is to maximize Z.

For the 3×2 example given above, Equations 5.1–5.6, this is

$$\begin{bmatrix} 10 & 5 \\ 1 & 1.5 \\ 2 & 2 \end{bmatrix} \begin{Bmatrix} X_c \\ X_b \end{Bmatrix} \le \begin{Bmatrix} W \\ L \\ F \end{Bmatrix} \tag{5.25}$$

$$\begin{Bmatrix} X_c \\ X_b \end{Bmatrix} \ge 0 \tag{5.26}$$

$$\begin{Bmatrix} V_c & V_b \end{Bmatrix} \begin{Bmatrix} X_c \\ X_b \end{Bmatrix} = Z \tag{5.27}$$

It is useful to formalize the *slack variables* S_j, defined as the unused portion of the
jth resource. These were recognized in Tables 5.2 and 5.4. Inserting these into the
resource constraints converts them to equalities:

$$\begin{bmatrix} 10 & 5 & | & 1 & 0 & 0 \\ 1 & 1.5 & | & 0 & 1 & 0 \\ 2 & 2 & | & 0 & 0 & 1 \end{bmatrix} \begin{Bmatrix} X_c \\ X_b \\ S_x \\ S_l \\ S_f \end{Bmatrix} = \begin{Bmatrix} W \\ L \\ F \end{Bmatrix} \tag{5.28}$$

$$\begin{Bmatrix} X_c \\ X_b \\ S_x \\ S_l \\ S_f \end{Bmatrix} \ge 0 \tag{5.29}$$

---

[2] Superscript ^{T} indicates transposition.

$$\left\{ \begin{array}{ccc|ccc} V_c & V_b & V_w & V_s & V_f \end{array} \right\} \left\{ \begin{array}{c} X_c \\ X_b \\ \hline S_x \\ S_l \\ S_f \end{array} \right\} = Z \tag{5.30}$$

The sense of inequality is transferred to the nonnegativity constraints on the slacks. The objective is generalized trivially by setting the values of the slacks V_w etc. to zero. (In a larger context, this provides some useful possibilities – for example, if the slack has value in itself.)

This *augmented form* is easily formalized:

$$[A']\{X'\} = \{R\} \tag{5.31}$$

$$\{X'\} \geq \{0\} \tag{5.32}$$

$$\{V'\}^T\{X'\} = Z \tag{5.33}$$

The prime indicates an augmented quantity. The unknown *decision variables* include the activities X and the resource slacks S. Specification of these decisions completely determines the *program of activity*. The available resource levels are external to the problem and appear as before on the right side, R. They constrain the possible programs. The objective is unchanged: to maximize value Z, a linear function of the decisions.

5.2.4 Networked Hydrology

Next, consider two riparian communities arranged as shown in Figure 5.3. To approach this situation, we need to distinguish hydrologically among river flows Q, withdrawal rates W, and return flows R. These are all *rates* (volume/time), and we assume the network is in quasistatic equilibrium. All of these flow rates are unknown, to be determined. The two exogenous constant inflows I are presumed known.

The network balance at each node of the network is

$$Q_1 = I_1 \tag{5.34}$$

$$Q_2 = Q_1 - W_1 \tag{5.35}$$

$$Q_3 = Q_2 + R_1 \tag{5.36}$$

$$Q_4 = Q_3 + I_2 \tag{5.37}$$

$$Q_5 = Q_4 - W_2 \tag{5.38}$$

$$Q_6 = Q_5 + R_2 \tag{5.39}$$

These are, of course, all linear, and they fit nicely into the LP format introduced above. Suppose that each withdrawal creates a proportional return flow:

$$R_i = \rho_i W_i \tag{5.40}$$

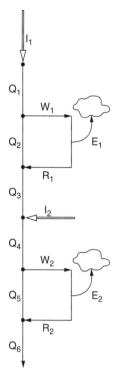

Figure 5.3. Two communities share the same river. There is an obvious direc-
tionality to the flow, hence an upstream/downstream distinction among the
riparians. Inflows I, in-channel flows Q, withdrawals W, consumptive use
(evapotranspiration) E, and return flows R are distinguished; they are all
nonnegative.

If W_i is withdrawn, then $(1 - \rho_i)W_i$ is consumed or converted to vapor, while the
remainder $\rho_i W_i$ is returned as liquid drainage to the network. In this case, we can
eliminate R_i and the linear relations reduce to

$$
\begin{bmatrix}
1 & 0 & 0 & 0 & 0 & 0 & 0 & 0 \\
-1 & 1 & 0 & 0 & 0 & 0 & 1 & 0 \\
0 & -1 & 1 & 0 & 0 & 0 & -\rho_1 & 0 \\
0 & 0 & -1 & 1 & 0 & 0 & 0 & 0 \\
0 & 0 & 0 & -1 & 1 & 0 & 0 & 1 \\
0 & 0 & 0 & 0 & -1 & 1 & 0 & -\rho_2
\end{bmatrix}
\begin{Bmatrix}
Q_1 \\ Q_2 \\ Q_3 \\ Q_4 \\ Q_5 \\ Q_6 \\ \hline W_1 \\ W_2
\end{Bmatrix}
=
\begin{Bmatrix}
I_1 \\ 0 \\ 0 \\ I_2 \\ 0 \\ 0
\end{Bmatrix}
\tag{5.41}
$$

We summarize this as the network continuity equation

$$[C']\{Q'\} = \{I\} \tag{5.42}$$

relating exogenous inflows I to their distribution among natural flows Q and withdrawals W. (The prime notation indicates that Q is augmented by W.) There is exactly
one constraint per network node.

All hydrological constraints defined here are strict equalities. There is no sense of
slack variables. And as binding equalities, they always have nonzero shadow prices.
However, as equalities, *the shadow prices may be either positive or negative.* One

way to imagine this is to conceive of the equality as simultaneously a positive and a negative constraint, one of which binds while the other is slack. Depending on the full set of constraints, increasing a flow may or may not be beneficial to the objective.

Implied in all of this is that the network flows are nonnegative:

$$\{Q'\} \geq \{0\} \tag{5.43}$$

A more general relation is useful when R is not simply proportional to W – for example, it may depend on just how a withdrawal is used. In that case, we need to retain R as variables and find closure relations outside the hydrology per se. The continuity equation in that more general case would be

$$
\begin{bmatrix}
1 & 0 & 0 & 0 & 0 & 0 & 0 & 0 & 0 & 0 \\
-1 & 1 & 0 & 0 & 0 & 0 & 1 & 0 & 0 & 0 \\
0 & -1 & 1 & 0 & 0 & 0 & 0 & -1 & 0 & 0 \\
0 & 0 & -1 & 1 & 0 & 0 & 0 & 0 & 0 & 0 \\
0 & 0 & 0 & -1 & 1 & 0 & 0 & 0 & 1 & 0 \\
0 & 0 & 0 & 0 & -1 & 1 & 0 & 0 & 0 & -1
\end{bmatrix}
\begin{Bmatrix}
Q_1 \\ Q_2 \\ Q_3 \\ Q_4 \\ Q_5 \\ Q_6 \\ W_1 \\ R_1 \\ W_2 \\ R_2
\end{Bmatrix}
=
\begin{Bmatrix}
I_1 \\ 0 \\ 0 \\ I_2 \\ 0 \\ 0
\end{Bmatrix}
\tag{5.44}
$$

This fits the same general continuity equation form as in Equation 5.42:

$$[C'']\{Q''\} = \{I\} \tag{5.45}$$

$$\{Q''\} \geq \{0\} \tag{5.46}$$

where the double primes indicate that Q is augmented by W and R.

Many solutions are possible that satisfy continuity.

Consumptive Use

W and R are not independent. Closing up this system of flows requires constitutive relations between them – essentially, the specification of what portion of the withdrawal is *consumed* (evaporated) and what part is returned to the stream. This requires knowing the activity X accompanying each withdrawal and its consumptive rate E:

$$R = W - E(X) \tag{5.47}$$

These *constitutive relations* are strict equality constraints. *One will be required for each withdrawal,* linking the hydrology to the economic activity. Any water not consumed (evaporated) *must* be returned to the stream. (These are energetic flows!)

5.2.5 **Hydroeconomy**

Suppose the two communities are both farms as described above. Significantly, the water resource available at either farm is not exogenous, but a decision variable intimately bound up in the basin hydrology. The economic supplement to the hydrology for Farm 1 would be analogous to Equation 5.25:

$$\begin{bmatrix} 10 & 5 & -1 \\ 1 & 1.5 & 0 \\ 2 & 2 & 0 \end{bmatrix} \begin{Bmatrix} X_c \\ X_b \\ W_1 \end{Bmatrix} \leq \begin{Bmatrix} 0 \\ L \\ F \end{Bmatrix} \tag{5.48}$$

The variable W_1 has migrated to the left side of this equation, in keeping with its internal, unknown status, leaving the other exogenous inputs L and F on the right side as input data. W_1 and the other hydrological variables are now part of the program, to be found.

We have yet to relate the return flow R to the activity level. Let e_i be the *consumptive demand* of water used by crop i (the amount converted to vapor), per unit of crop. Then we have the requirement

$$e_c X_c + e_b X_b + R_1 = W_1 \tag{5.49}$$

Essentially, cotton requires withdrawal of 10 units of water per unit grown; $e_c \leq 10$ is the amount of it that is actually consumed by the cotton. What does not get consumed returns to the stream either directly through drainage or indirectly through groundwater recharge. The complete relation for this farm is then

$$\begin{bmatrix} 10 & 5 & -1 & 0 \\ 1 & 1.5 & 0 & 0 \\ 2 & 2 & 0 & 0 \\ e_c & e_b & -1 & +1 \end{bmatrix} \begin{Bmatrix} X_c \\ X_b \\ W_1 \\ R_1 \end{Bmatrix} \sim \begin{Bmatrix} \leq 0 \\ \leq L \\ \leq F \\ = 0 \end{Bmatrix} \tag{5.50}$$

Here the $\sim$ indicates "is required to be"; the relational sense of each scalar equation is moved inside the right-side vector.

Notice the important distinction between the minimum *withdrawal* required by an activity (the first constraint) and the water actually *consumed* by that activity (the last equality). Activity will normally require $W > E$; the limiting nonconsumptive use still requires W, thereby competing with opportunities in the reaches upstream of the return flow R.

A similar representation of activity at the downstream farm would involve the withdrawal W_2 as an endogenous variable input to the production and the return flow R_2 as an outcome dependent on production. Naturally, all of the continuity requirements defined above are assumed to govern the network of flows.

Suppose the two farms are growing the same two crops with the same technology; the hydrology is as given above; the land availability is L_1 and L_2; and the fertilizer resource is F_1 and F_2. The agricultural production constraints would be the simple concatenation

$$
\begin{bmatrix}
10 & 5 & 0 & 0 & -1 & 0 & 0 & 0 \\
1 & 1.5 & 0 & 0 & 0 & 0 & 0 & 0 \\
2 & 2 & 0 & 0 & 0 & 0 & 0 & 0 \\
e_c & e_b & 0 & 0 & -1 & 1 & 0 & 0 \\
0 & 0 & 10 & 5 & 0 & 0 & -1 & 0 \\
0 & 0 & 1 & 1.5 & 0 & 0 & 0 & 0 \\
0 & 0 & 2 & 2 & 0 & 0 & 0 & 0 \\
0 & 0 & e_c & e_b & 0 & 0 & -1 & 1
\end{bmatrix}
\left\{
\begin{array}{c}
X_{c1} \\ X_{b1} \\ X_{c2} \\ X_{b2} \\ W_1 \\ R_1 \\ W_2 \\ R_2
\end{array}
\right\}
\sim
\left\{
\begin{array}{c}
\leq 0 \\ \leq L_1 \\ \leq F_1 \\ = 0 \\ \leq 0 \\ \leq L_2 \\ \leq F_2 \\ = 0
\end{array}
\right\}
\tag{5.51}
$$

Suppose the the fertilizer resource $F = F_1 + F_2$ is shared among them. Then the two fertilizer constraints merge:

$$
\begin{bmatrix}
10 & 5 & 0 & 0 & -1 & 0 & 0 & 0 \\
1 & 1.5 & 0 & 0 & 0 & 0 & 0 & 0 \\
e_c & e_b & 0 & 0 & -1 & 1 & 0 & 0 \\
2 & 2 & 2 & 2 & 0 & 0 & 0 & 0 \\
0 & 0 & 10 & 5 & 0 & 0 & -1 & 0 \\
0 & 0 & 1 & 1.5 & 0 & 0 & 0 & 0 \\
0 & 0 & e_c & e_b & 0 & 0 & -1 & 1
\end{bmatrix}
\left\{
\begin{array}{c}
X_{c1} \\ X_{b1} \\ X_{c2} \\ X_{b2} \\ W_1 \\ R_1 \\ W_2 \\ R_2
\end{array}
\right\}
\sim
\left\{
\begin{array}{c}
\leq 0 \\ \leq L_1 \\ = 0 \\ \leq F \\ \leq 0 \\ \leq L_2 \\ = 0
\end{array}
\right\}
\tag{5.52}
$$

In each case here, the hydrological constraints (continuity equation, in the form of Equation 5.44) are assumed to be appended, and all decision variables are required to be nonnegative.

The program **TwoFarms.xls** implements this optimization in the Excel Solver.

5.2.6 Municipal Water Supply

Suppose the downstream community is instead a city, with service population P_2. There will be a water supply requirement that scales with population:

$$W_2 \geq w_p P_2 \tag{5.53}$$

and return flow will be generated with consumptive use a by-product of service and proportional to the service population

$$E_2 = e_p P_2 \tag{5.54}$$

$$e_p P_2 + R_2 = W_2 \tag{5.55}$$

Finally, the population will have a land use requirement:

$$L_2 \geq h_p P_2 \tag{5.56}$$

Any of these can limit the population size, in concert with the other constraints already expressed.

The constitutive relation for this community is then

$$
\begin{bmatrix}
w_p & -1 & 0 \\
h_p & 0 & 0 \\
e_p & -1 & 1
\end{bmatrix}
\begin{Bmatrix}
P_2 \\
W_2 \\
R_2
\end{Bmatrix}
\sim
\begin{Bmatrix}
\leq 0 \\
\leq L_2 \\
= 0
\end{Bmatrix}
\tag{5.57}
$$

The cropping activities have been removed; in their place, the service population P becomes a decision variable.

Coupling this city with the upstream farm gives the composite set of relations

$$
\left[
\begin{array}{cc|c|cc|cc}
10 & 5 & 0 & -1 & 0 & 0 & 0 \\
1 & 1.5 & 0 & 0 & 0 & 0 & 0 \\
2 & 2 & 0 & 0 & 0 & 0 & 0 \\
e_c & e_b & 0 & -1 & 1 & 0 & 0 \\
\hline
0 & 0 & w_p & 0 & 0 & -1 & 0 \\
0 & 0 & h_p & 0 & 0 & 0 & 0 \\
0 & 0 & e_p & 0 & 0 & -1 & 1
\end{array}
\right]
\left\{
\begin{array}{c}
X_{c1} \\
X_{b1} \\
\hline
P_2 \\
\hline
W_1 \\
\hline
R_1 \\
W_2 \\
R_2
\end{array}
\right\}
\sim
\left\{
\begin{array}{c}
\leq 0 \\
\leq L_1 \\
\leq F_1 \\
= 0 \\
\leq 0 \\
\leq L_2 \\
= 0
\end{array}
\right\}
\tag{5.58}
$$

Of course, these constitutive relations complement the general continuity equations that were introduced earlier to enforce the network topology.

5.2.7 Hydropower

Suppose hydraulic machinery can be arranged such that a flow rate Q (volume/time) passes through it nonconsumptively, experiencing a pressure drop ΔP. The power generated $\dot{E}$ (energy/time) will be the product $\rho \, Q \, \Delta P$ (here, ρ is the fluid density). We are concerned here with free surface flows, so the fluid is hydrostatic and the pressure drop will be proportional to the elevation change, H ("head"). In MKS units, for water at ambient conditions: if $[Q] = \mathrm{m}^3/\mathrm{s}$, $[H] = \mathrm{m}$, then the power $\dot{E}$ in kilowatts is

$$
\dot{E} = 9.81 \, \epsilon \, Q H \tag{5.59}
$$

where ϵ is the efficiency of conversion from mechanical to electrical energy. Energy produced would be the time integral of $\dot{E}$ in Equation 5.59.

As we are concerned with flow rates, much will depend on the head H. There are two extremes. In the simple case, H is fixed for a given installation. Essentially, there is no fluctuation of water level behind the intake for the power machinery. In this case, power is proportional to flow rate Q – a linear relation. "Run-of-River" installations are close to this; they typically operate with little or no change in H. In periods of high Q, the flow that cannot be accommodated by the installed machinery, bypasses it, spilling "over the dam" or around it.

In the other extreme, storage reservoirs are built behind major dams, with the intention to absorb incoming flow variability in storage fluctuations, thereby making possible less variable withdrawals and downstream flows. Power produced in these facilities is fundamentally nonlinear. The topography of the reservoir and dam combination results in a static relation between water storage S and the height of the free surface H, the "storage-head relation":

$$H = H(S) \qquad (5.60)$$

Power–law relations are common approximations, $H = S^{\alpha}$, with $\alpha < 1$ accounting for increasing surface area as S increases. There will be more on this in later sections.

5.2.8 Navigation

Navigational use will normally be conceived of as nonconsumptive, relying only on the maintenance of flat water (as in an impoundment or lake) and/or flow rate through steeper channels. In the extreme, locks are required for vessel transit. Any given installation will require the flow of a certain flow rate per vessel and impose a ceiling on the number of vessels that may transit. This use is therefore tantamount to maintaining certain flow rates in navigable network links in the steady state. If a network is not in steady operation, then the storage levels as well as the flows will be constrained by navigational use. If a toll is collectible, income will be proportional to the number of vessels transiting a facility.

Fish ladders can be considered analogously: nonconsumptive flow required for transit. It is unlikely that there is any analogue of the maximum transit capacity, however – simply a requirement for flow rate in a constructed facility.

5.3 EXAMPLE: THE WENTWORTH BASIN

We will consider a comprehensive case study in the Wentworth Basin. Its hydrological network is shown in Figure 5.4.

Two farms are arranged in this basin as shown. Each requires the blends of water, land, fertilizer, and labor given in Table 5.5, per unit of production. Farm 1 has 5 units of land; Farm 2 has 10. Fertilizer may be shared among the two farms; the total availability is 12 units. Labor is assumed to be in abundant supply. Consumptive use at each farm is assumed to be 0.6 of the withdrawal.

Water supply is limited by the hydrologic network. Of course, the continuity equation governs; in addition, there are two extra policy constraints:

- Flow must be maintained in all reaches of the river at the level 0.2 or higher.
- Flow exiting the country is required by treaty to be 50% of the total natural flow (half of $1.25 + .75$).

The objective is to maximize the value of activity. The Excel programs **Wentworth1** through **Wentworth6** are used to solve Cases 1–6.

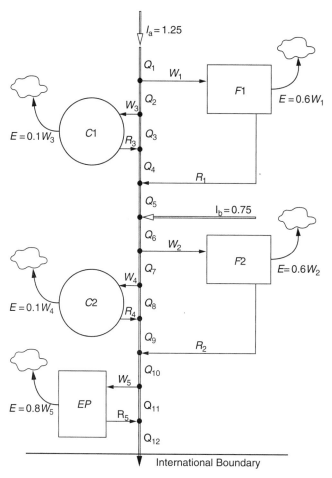

Figure 5.4. Wentworth Basin with two farm regions, two residential communities, an ethanol plant, and a downstream riparian nation. Consumptive flows $E = (1 - \rho)W$ are indicated.

Table 5.5. Crop requirements for the two farms in the Wentworth Basin per unit of production. There are four inputs to production for each crop.

	Farm 1		Farm 2	
Inputs	Cotton	Wheat	Wheat	Corn
Land	5	4	4	5
Water	2	8	8	5
Fertilizer	1	1	1	1
Labor	2	3	3	4
Value	$10	$15	$15	$20
Crop amount	X_{c1}	X_{w1}	X_{w2}	X_{c2}

Table 5.6. Results of the Wentworth optimization, by case.

				Case			
* Program:	1	2	3	4*	5	6**	6a***
Cotton$_1$	.525	.5	.24	0	0	0	0
Wheat$_1$	0	0	.068	.003	0	.1235	0
Wheat$_2$	0	0	.038	.094	0	.0827	.2062
Corn$_2$	.123	.131	.067	.165	0	0	0
Labor$_1$	1.05	1	.677	.029	0	1.235	0
Labor$_2$	.492	.523	.383	.94	0	.827	2.062
Ethanol	0	0	0	.0413	0	0	0
Power	0	0	0	0	$20	0	0
Money	$7.717	$7.27	$3.47	$4.733*	$20	−$0.464	−$0.464
				($0.60)			
W_1	1.05	1	1.02	.024	0	.988	0
W_2	.617	.654	.64	1.58	0	.661	1.649
W_3	0	.05	.034	.0015	0	.062	0
W_4	0	.026	.019	.047	0	.041	.1031
W_5	0	0	0	.041	0	0	0
Q_1	1.25	1.25	1.25	1.25	1.25	1.25	1.25
Q_2	.2	.25	.234	1.226	1.25	.262	1.25
Q_3	.2	.2	.2	1.225	1.25	.2	1.25
Q_4	.2	.245	.230	1.226	1.25	.256	1.25
Q_5	.62	.645	.637	1.236	1.25	.651	1.25
Q_6	1.37	1.395	1.39	1.986	2	1.40	2
Q_7	.753	.741	.745	.406	2	.740	.351
Q_8	.753	.715	.726	.359	2	.698	.247
Q_9	.753	.738	.743	.401	2	.736	.340
Q_{10}	1	1	1	1.033	2	1	1
Q_{11}	1	1	1	.992	2	1	1
Q_{12}	1	1	1	1	2	1	1

*Case 4 is computed with the subsidy $S = 100$; the "profit" being maximized, 4.733, is exaggerated by the amount $S \times E$. Accounting for that leaves a net profit of $0.60. A subsidy less than $60 has no effect, resulting in the same optimal program as in Case 3. **Case 6 is the same as Case 3, with the objective changed to maximize employment and ignore money. ***Cases 6 and 6a are equivalent, reflecting multiple optima.

Case 1. This is the base case. All of the needed constraint forms have been reviewed above; they are implemented in the program **Wentworth1.xls**. Results appear in Table 5.6. There is no wheat produced; labor is unconstrained and so has no effect on the program. The only withdrawals are for the two farms where value can be created. The minimum flows bind in Reaches 2, 3, and 4. The treaty requirement constrains the lower half of the basin.

Case 2. To the base case, add a requirement for public health as follows: It is now recognized that people need to be employed on the farm from the nearest adjacent cities, as shown. The laboring population L_1 and L_2 are unknown; but domestic water supply is needed in the cities at the rate $W > .05L$ in each case. About 90% of this withdrawal will be returned to the stream after treatment; the mandatory treatment cost is $5 per unit of flow returned.

In this case, the following additions to the formulation from Case 1 are needed:

- L_1 and L_2 need to be added as nonnegative decision variables.
- The farm constraints on labor need to be added:

$$2X_{c1} + 3X_{w1} - L_1 \leq 0 \tag{5.61}$$

$$3X_{w2} + 4X_{c2} - L_2 \leq 0 \tag{5.62}$$

- The public health constraints need to be added:

$$W_3 - .05L_1 \geq 0 \tag{5.63}$$

$$W_4 - .05L_2 \geq 0 \tag{5.64}$$

- The cost of pollution control needs to be recognized. From the Case 1 objective, subtract its cost:

$$Z \rightarrow Z - 5(.9W_3 + .9W_4) \tag{5.65}$$

Results appear in Table 5.6 as Case 2. There is a different crop mix, with fewer people employed upstream, where water is short. Withdrawals are evident for both cities. The minimum flow continues to bind at Reach 3, just downstream of the withdrawal for City 1. The treaty requirement still constrains the lower half of the basin. There is less money being made; that plus the small shift in the crop patterns accompanies the increase in public health provision through sanitation.

Case 3. Now suppose we add the requirement that some of the local agricultural production will be consumed locally. Effectively, each unit of labor requires 0.1 unit of local wheat that is produced but consumed locally – hence, not sold for profit.

Modifications to Case 2 include

- Wheat production must exceed that required by labor:

$$X_{w1} - .1L_1 \geq 0 \tag{5.66}$$

$$X_{w2} - .1L_2 \geq 0 \tag{5.67}$$

- The financial value of the program needs to be decremented for the wheat not sold on the open market (in Cases 1 and 2, *all* the wheat was credited as sold):

$$Z \rightarrow Z - 15(.1L_1 + .1L_2) \tag{5.68}$$

Results for this case are quite different from those above. Cotton and corn production are roughly halved, and wheat production is positive. Earnings are significantly reduced, reflecting local consumption. Labor is reduced, and so the urban flows are reduced, too. The minimum flow in Reach 3 continues to bind, as does the treaty requirement in the lower basin.

Case 4. Some or all of the corn grown on Farm 2 may now be used in an ethanol plant downstream, as shown. Each unit of fuel produced requires the following resource requirements: water, 1.0; corn, 4.0; labor, 0.5. The labor may reside in either city. The value of a unit of fuel is $20. The sale price will be subsidized by an amount S, below, but to start $S = 0$.

This case requires several modifications to the Case 3 formulation:

- New nonnegative decision variables are needed for ethanol production E and labor in the various cities devoted to ethanol, L_{E1} and L_{E2}. We consider L_1 and L_2 to be the labor *housed* in Cities 1 and 2, respectively; but now only a portion of it works on the farms:

$$L_{E1} - L_1 \leq 0 \tag{5.69}$$

$$L_{E2} - L_2 \leq 0 \tag{5.70}$$

- The farm labor constraints need to recognize this:

$$2X_{c1} + 3X_{w1} - (L_1 - L_{E1}) \leq 0 \tag{5.71}$$

$$3X_{w2} + 4X_{c2} - (L_2 - L_{E2}) \leq 0 \tag{5.72}$$

- Ethanol production needs to be constrained by its resources (water, corn produced, and labor):

$$1E - 1W_5 \leq 0 \tag{5.73}$$

$$4E - 1X_{c2} \leq 0 \tag{5.74}$$

$$.5E - (L_{E1} + L_{E2}) \leq 0 \tag{5.75}$$

- The financial value needs to be adjusted to reflect (a) the value of ethanol; and (b) the fact that some corn already accounted for as "sold", will not be brought to market, but rather embedded in the ethanol:

$$Z \to Z + (20 + S)E - 20(4E) = Z + (S - 60)E \tag{5.76}$$

The net value of a unit of fuel is $S - 60$. Clearly, $S < 60$ results in no ethanol production. A subsidy $S > 60$ makes ethanol a possibility. Table 5.6 displays this case with $S = 100$ as Case 4. At this level of subsidy, the basin is specializing in corn production downstream, nearly closing out the upstream economy. The profit is increased, although it is not accounting for the cost of the subsidy, which would be $100E$, hence a net profit of roughly $0.60. It is the subsidized profit that is maximized in this case. If the real profit were optimized, properly accounting for the subsidy, the optimal solution would revert to Case 3 with no ethanol production.

Case 5. Returning to Case 4 with zero subsidy, suppose that power could be produced at a rapids at C_2 (Q_8), under run-of-river operation. Essentially, power is produced at the value $10 per unit of flow there:

$$Z \to Z + 10Q_8 \tag{5.77}$$

This has the effect of shutting the whole basin down, preserving all flow for power production just before it enters the downstream state. The treaty is unnecessary because the most valuable use is nonconsumptive and no consumption is possible downstream of the rapids at Q_8, except for the unsubsidized ethanol plant.

This basin configuration is one of total export. Both water and power are harvested and exported nonconsumptively. The resource supports no population or local economy – there is zero in-basin development. This illustrates a subtle effect of ownership and decision control: If the objective is simply to make money, the result is to shut the Wentworth Basin down and export its water and power to financially attractive external customers.

Case 6. Using the conditions of Case 3, change the objective: Maximize employment $(L_1 + L_2)$ and ignore the money. Although all the constraints of Case 3 are left intact, the optimal program is very different. The requirement of local wheat consumption causes the solution to select this crop alone and to shut down the cash crops cotton and corn. Labor is approximately doubled. The money becomes negative,[3] reflecting (a) no cash crops; and (b) real pollution control costs. The latter cost would require a cash subsidy to the region. The minimum flow rate at Q_3 constrains upstream population; the treaty flow rate constrains downstream activity.

In fact, there are multiple optima for this case. Wheat production at both farms is identical, as are the conditions in both cities. The labor can therefore be reapportioned to the downstream farm and city, with the same total consumptive use. A limiting case would be shutting the upstream farm and city entirely. In that case, the minimum flow is maintained slack everywhere, but the whole location remains constrained by the border treaty. Total labor is the same, 2.062, with $L_1 = 0$. The money, although unconstrained, remains the same: the cost of pollution control for the cities. (For problems with multiple optima, the LP solver results would typically depend on details of starting conditions and/or numerical parameters.) This is Case 6a in Table 5.6.

5.4 INTEGERS

So far, we have concentrated on continuous decision variables: the hydrology Q, W, and R; the economy X; and the metrics of achievement Z. Suppose there are additional discrete questions – characterized by integer variables. These will provide

---

[3] Money is not required to be nonnegative in this example.

very important opportunities for formulation and implementation. To distinguish these, we will denote them by the variables Y_i, the ith integer variable. We will concentrate on the binary integers – those having only two values: 0 or 1. Many important decisions are binary themselves; and more generally, all other integers can be built up as a collection of binaries (i.e., can be represented in base 2).

The apparatus of binary-integer programming is arranged to accommodate this complexity. It requires constraints and objective to be linear in the decisions, as in LP; but the decisions themselves are binary integers, not real numbers. This changes the mathematical nature of the problem, the associated algorithms, and the properties of the solution. The generalization mixed-integer programming (MIP) employs a mix of binary and real decisions. The canonical formulation is the same as the standard LP formulation – linear constraints, linear objective – but the identities of the variables are distinguished. So all MIP formulations need to look like standard LP formulations; we will distinguish the binary variables by the symbols Y_i. Contemporary solution packages can be expected to accommodate the MIP case. (The Excel Solver does this.) This represents a significant broadening of the LP capability, and optimization packages that lack this feature are accordingly more restrictive.

5.4.1 Capital Cost

Let's reconsider the Wentworth Basin case (Case 2 above), where we were concerned with public health in the cities. It would be common to represent the cost of the water treatment in two parts: one, build it; and two, operate it. The building cost would include necessary design, permitting, land use permitting, sewerage and machinery. This would be a capital cost; it would be avoided if there were to be no use, but any use would be logically posterior. Hence, we have two costs: K, the annual value of the capital cost; and C, the annual cost per unit of treatment. Related would be two decision variables: the binary Y indicating the strategic build/no build decision; and the real nonnegative return flow R requiring treatment. The formulation for the treatment cost is

$$Z \to Z - KY - CR \tag{5.78}$$

$$R \leq YR^* \tag{5.79}$$

The datum R^* would be the capacity of the treatment facility covered by the capital cost K. If $Y = 1$, the plant is built and paid for, and R is enabled up to the plant capacity. Otherwise, $Y = 0$, there is no R permitted, and K is not assessed.

For a single decision, we have introduced two data related to the constuction (capital cost K and capacity R^*) and one binary variable. There would be two possibilities – build or not build – and these *could* be easily studied and optimized separately and the better option chosen without the extra integer apparatus. The outcome would be the same as if the integer formulation were used. This simpler approach, however, loses appeal as the problem grows in size and complexity.

In the Wentworth Basin example, there are two cities producing R_3 and R_4 and thus two binaries:

$$Z \rightarrow Z - K_3Y_3 - K_4Y_4 - C_3R_3 - C_4R_4 \tag{5.80}$$

$$R_3 - R_3^*Y_3 \leq 0 \tag{5.81}$$

$$R_4 - R_4^*Y_4 \leq 0 \tag{5.82}$$

Now there are four distinct possibilities, four separate optima, and a comparison is needed among the four. This rapidly scales up in complexity: N binary decisions will generate 2^N distinct possibilities. The automation of this via MIP is a distinct advantage.

5.4.2 Sequencing

Suppose there were a possibility of adding capacity to the downstream plant, adding R_{4a}^* to the capacity at extra cost K_{4a}. Let the decision variable here be Y_{4a}. This decision is contingent on the plant existing at all, that is, on the decision Y_4 having been taken. The formulation would resemble a 3-facility decision, with a sequencing contingency added:

$$Z \rightarrow Z - K_3Y_3 - K_4Y_4 - K_{4a}Y_{4a} - C_3R_3 - C_4R_4 \tag{5.83}$$

$$R_3 - R_3^*Y_3 \leq 0 \tag{5.84}$$

$$R_4 - R_4^*Y_4 - R_{4a}^*Y_{4a} \leq 0 \tag{5.85}$$

$$Y_{4a} \leq Y_4 \tag{5.86}$$

5.4.3 Alternative Constraints

Suppose the minimum flow in Reach 5 were under study, with three different values supporting three different river classifications: $Q_5 \geq .2$ or $.3$ or $.5$. Introduce three binaries, and require *exactly one* of them to be in force:

$$Q_5 - .2Y_b \geq 0 \tag{5.87}$$

$$Q_5 - .4Y_c \geq 0 \tag{5.88}$$

$$Q_5 - .5Y_d \geq 0 \tag{5.89}$$

$$Y_b + Y_c + Y_d = 1 \tag{5.90}$$

If *at most* one were required, the final constraint would become

$$Y_b + Y_c + Y_d \leq 1 \tag{5.91}$$

5.4.4 Interbasin Transfer

Suppose it were possible to import a flow from an adjacent basin, supplementing the natural flow $Q_b = .75$. Name this additional flow Q_c. The network would then have the modified continuity relation

$$Q_b - Q_c = .75 \tag{5.92}$$

Q_c would require a capital investment K cost related to right-of-way, withdrawal permitting at the source, and the construction of conveyance facilities. This would buy a capacity of Q_c^*. In addition, it would require pumping costs C per unit of flow. The integer formulation would then use the apparatus of capital cost introduced above:

$$Z \rightarrow Z - KY_c - CQ_c \tag{5.93}$$

$$Q_c - Q_c^* Y_c \leq 0 \tag{5.94}$$

and of course the implied relations Y_c binary, Q_c real and nonnegative.

5.5 GOALS

The above development in terms of optimization suggests a unilateral devotion to a single goal, which is to be optimized. This is very unrealistic. Nearly every problem of interest has multiple, conflicting goals. We must not lose sight of this basic feature.

In the Wentworth Basin example, we summarized several solutions in which money was optimized. The money focused on the salable products (crops, power, and fuel) and accountable costs. But in each case, there are other outcomes associated with any particular program: the employed population; the public health provision in terms of water supply and wastewater requirements; the nutrition levels; the minimum flows; the provision of fish ladders and respect for the treaty. The case that maximized employment (rather than money) highlighted this by producing a very different optimal program. Each and every one of the cases studied is a possible program that *achieves the constraints* and, beyond that, *optimizes something* (e.g., cash or jobs in the Wentworth cases). So it is best to view these optima as solutions with many outcomes. A constraint represents a minimum level of achievement in resource management. The optimization proceeds to make the most of the remaining freedom in terms of a single objective. One could always constrain $N - 1$ of the important outcomes and optimize the final outcome. Each would be a suboptimization with N outcomes, and there would be N such suboptima.

5.5.1 Multiple Objectives

Suppose there are only four decision variables: Q, W, R and X. We could express several valuable objectives Z_i as linear combinations of these variables:

$$Z_i = V_{iQ}Q + V_{iW}W + V_{iR}R + V_{iX}X \tag{5.95}$$

where the subscript i indicates "objective i." Each can be considered a linear equality constraint, defining Z_i. In matrix form,

$$\begin{Bmatrix} Z_1 \\ Z_2 \\ \vdots \\ \vdots \\ Z_N \end{Bmatrix} - \begin{bmatrix} V_{1Q} & V_{1W} & V_{1R} & V_{1X} \\ V_{2Q} & V_{2W} & V_{2R} & V_{2X} \\ \vdots & \vdots & \vdots & \vdots \\ \vdots & \vdots & \vdots & \vdots \\ V_{NQ} & V_{NW} & V_{NR} & V_{NX} \end{bmatrix} \begin{Bmatrix} Q \\ W \\ R \\ X \end{Bmatrix} = \begin{Bmatrix} 0 \\ 0 \\ \vdots \\ \vdots \\ 0 \end{Bmatrix} \tag{5.96}$$

More compactly, for a vector of decision variables D,

$$\{Z\} - [V]\{D\} = \{0\} \tag{5.97}$$

(Of course, there could be more than one Q, etc. in D.) The value coefficients V_{ij} are "the value to objective i, of a unit of decision j." We have defined N new decision variables and added N equality constraints. Z_i would not naturally be required to be nonnegative; if desirable, that would require an additional constraint.

Following this, we move the objective into the constraints. Now a single-objective optimization can be stated simply. For example,

$$\max(Z_1) \tag{5.98}$$

or for a linear combination of objectives:

$$\max(Z_1 + 3Z_2 + .5Z_3) \tag{5.99}$$

Other possibilities include constraining certain objectives while maximizing others – for example,

$$\max(3Z_2 + .5Z_3) \tag{5.100}$$

subject to the constraint

$$Z_1 \geq 10 \tag{5.101}$$

We quickly open the way to the exploration of different metrics of value within the same set of constraints.

5.5.2 Metrics

The simplest, *Composite metric* is simply a weighted sum of the individual Z_i:

$$\max\left(\sum W_i Z_i = \{W\}^T \{Z\}\right) \tag{5.102}$$

as illustrated in Equation 5.100. The W_i are often thought of as representing political influence; the criterion is mathematically the same as that representing economic values. It would be common to supplement these with a requirement of minimum levels of achievement L_i for some or all of the Z_i:

$$\{Z\} \geq \{L\} \tag{5.103}$$

In other cases, we may be concerned with a *Rawlsian metric*: the least among many goals, $\min_i (Z_i)$, without regard to which one it is. The related criterion would be to make this smallest Z_i as high as possible, saying nothing about the sizes (or relative sizes) of the others. This would lead to the well-known MaxMin criterion:

$$\max \left(\min_i (Z_i) \right) \tag{5.104}$$

This criterion is readily implemented in LP form. First, define an unknown lower bound as the decision variable $\underline{Z}$:

$$\underline{Z} - Z_i \leq 0 \tag{5.105}$$

for all i. Then make the objective

$$\max(\underline{Z}) \tag{5.106}$$

The analogous problem is the minmax, wherein we define $\overline{Z}$ the highest of the Z_i:

$$\overline{Z} - Z_i \geq 0 \tag{5.107}$$

for all i. Then make the objective

$$\min(\overline{Z}) \tag{5.108}$$

Yet other goals might be related to the achievement gaps among the Z_i. If we define the *Inequity metric* as the $\overline{Z}$ and $\underline{Z}$ as above, then inequity minimization would be

$$\min(\overline{Z} - \underline{Z}) \tag{5.109}$$

This would have to be further constrained with minimum achievements as in, for example, Equation 5.103, lest it result in all Z_i being zero: Do nothing and share the nonachievement equitably!

A better metric would focus on the *relative inequity*: Require the inequity metric to be proportional to the mean achievement. For example, for 30% we would have the constraint

$$\overline{Z} - \underline{Z} \leq \frac{0.3}{N} \sum_i Z_i \tag{5.110}$$

leaving the criterion to be maximized, to be specified in terms of the Z_i – For example, maximize the mean Z_i subject to the constraint equation 5.110 would take the form

$$\max \sum_i Z_i \tag{5.111}$$

subject to $2N + 1$ constraints

$$\overline{Z} - Z_i \geq 0 \qquad (5.112)$$

$$\underline{Z} - Z_i \leq 0 \qquad (5.113)$$

$$\overline{Z} - \underline{Z} - \frac{0.3}{N} \sum_i Z_i \leq 0 \qquad (5.114)$$

$$(5.115)$$

All of these amount to standard linear-programming formulations, implementable within the canonical LP format of linear inequalities and linear objectives being optimized.

5.5.3 Targets

It is not uncommon to have metrics that are assymetric, where underachievement is viewed differently than overachievement. This achievement has thus a central target value (data) and two *nonnegative* metrics of achievement Z_i^+ and Z_i^-:

$$Z_i \equiv T_i + Z_i^+ - Z_i^- \qquad (5.116)$$

The objectives are now

$$[V]\{D\} - \{Z\}^+ + \{Z\}^- = \{T\} \qquad (5.117)$$

To resolve indeterminacy, at least one of the two achievement variables must be zero. Let Y_i be a binary integer and M be an arbitrarily large number:

$$Z_i^+ \leq Y_i M \qquad (5.118)$$

$$Z_i^- \leq (1 - Y_i)M \qquad (5.119)$$

Hence, we achieve the assymetry of goals by introducing one datum T and three decision variables – two nonnegative real numbers and one binary integer – in place of the original decision variable Z_i. The resulting optimization can be stated in terms of the value of being off-target relative to the various goals; and constraints can be exercised on these goal discrepancies as well.

Example: A newly seated government has run on a platform of human rights, citing the Universal Declaration. There are four decision variables, one of which is not independent:

X_e: invest in education
W: increase the minimum wage
T: tax corporate earnings
G: economic growth, $G = -W + .5T$

Three key concerns are increases in employment E, personal income P, and educational achievement A. They relate to the decision variables as follows:

$$E = 0X_e - 1W - .5T + 2G \tag{5.120}$$

$$P = 2X_e + 2W + 0T + 1G \tag{5.121}$$

$$A = 8X_e + 0W + 2T + 1G \tag{5.122}$$

Advisers have decided on these goals:

- *Employment*: Increase by 50 is absolutely necessary; extra increase is valued at 1.
- *Income*: Increase by 10. Failure is acceptable but penalized by two per unit of underachievement. There is no value in exceeding this goal.
- *Education*: Increase by 20. This is absolutely necessary; and because it represents universal education, it cannot be exceeded.

The linear goal formulation is

$$0X_e - 1W - .5T + 2G - E^+ + E^- = 50 \tag{5.123}$$

$$2X_e + 2W + 0T + 1G - P^+ + P^- = 10 \tag{5.124}$$

$$8X_e + 0W + 2T + 1G = 20 \tag{5.125}$$

$$E^+ - Y_E M \le 0 \tag{5.126}$$

$$E^- - (1 - Y_E)M \le 0 \tag{5.127}$$

$$P^+ - Y_P M \le 0 \tag{5.128}$$

$$P^- - (1 - Y_P)M \le 0 \tag{5.129}$$

and the objective would be

$$\max(E^+ - ME^- + 0P^+ - 2P^-) \tag{5.130}$$

As above, decision variables $E^{\pm}$, $P^{\pm}$ are nonnegative; the binary integers Y are introduced to avoid indeterminacy in E and P; and M is an arbitrarily large number. (Equivalently, we could in this case simply eliminate E^- and Y_E from the formulation entirely – less systematic, but simpler for this case.)

5.5.4 Regret

Suppose we know the highest level of achievement possible for any objective, Z_i^*. (This could be found by a simple maximization of Z_i alone and then treating the outcome Z_i^* as data.) Clearly, any other outcome subject to the same constraints will be inferior, and the distance is the regret R_i:

$$R_i \equiv Z_i^* - Z_i \tag{5.131}$$

Regret, so defined, will always be nonnegative. Minimizing R_i would be the same as maximizing Z_i. If Z measures achievement, R measures (regrettable) underachievement.

With multiple Z_i, some regret is unavoidable. We could devise regret metrics in the same way as above for Z: various combinations of R_i and/or constraints on them. Any metric that is linear in the Z_i will be linear in the R_i, and maximizing one is the same as minimizing the other. However, this equivalence does not hold if the objective is nonlinear in R_i.

The extreme case is the Rawlsian or minmax criterion: Minimize the worst-case regret among the i without knowing which i is the worst. This involves defining a maximum regret R^* and minimizing it:

$$\min R^* \tag{5.132}$$

subject to

$$R^* \geq R_i \tag{5.133}$$

or equivalently,

$$R^* + Z_i \geq Z_i^* \tag{5.134}$$

This formulation has the twin charactistics of (a) focusing on underachievement of independent Z_i; and (b) guaranteeing a ceiling on all, without preference among them. It requires finding Z_i^* first by N narrowly focused optimizations and then using each result as data in a final minmax optimization. There are therefore $N + 1$ separate optimizations, all sharing identical constraints and differing only in the specific objective.

5.5.5 Merit

In Equation 5.97, we identified all objectives Z_i as linear combinations of the decision variables D:

$$\{Z\} = [V]\{D\} \tag{5.135}$$

The D comprise flows, withdrawals, economic activities, etc. This would be a standard utilitarian formulation: individual satisfaction as a linear combination of material decisions.

Suppose Z is supplemented by an additional *merit* term reflecting satisfaction with another's happiness:

$$\{Z\} = [V]\{D\} + [M]\{Z\} \tag{5.136}$$

The merit matrix $[M]$ would have zeros on the diagonals and nonzero off-diagonal terms.[4] Individual satisfaction now recognizes two terms: direct (V) and social (M).

---

[4] The zero-diagonal property is necessary to resolve ambiguity between $[V]$ and $[M]$.

Rearrangement gives

$$[I - M]\{Z\} = [V]\{D\} \tag{5.137}$$

$$\{Z\} = [I - M]^{-1}[V]\{D\} \tag{5.138}$$

We can identify a *solidarity* matrix $[S] \equiv [I - M]^{-1}$, conditioning the utilitarian matrix $[V]$:

$$\{Z\} = [S][V]\{D\} \tag{5.139}$$

This formulation is in the same form as the simpler Equation 5.135 – Z a linear transformation of D. Its foundation reflects important distinctions, however. For example, in quantifying the coefficients, one needs to imagine an experiment of changing one item – for example, D_j – and measuring the set of changes in Z; *while keeping all else constant.* In this perturbation, the distinction between the two partial derivatives – holding D_k constant and holding Z_l constant – is substantive. Related is observability and the role of error in measuring.

There is a wide and controversial literature surrounding merit goods (Musgrave and Musgrave 1989; Ver Eecke 2007). Here we seek only to introduce the topic and invite scrutiny. The issues are both philosophical, at the foundation of economic theory; and operational, affecting the interpretation of economic data and behavior.

5.6 DYNAMICS

So far, we have assumed a steady network. If we relax this requirement, then there must be storage $S_i(t)$ accounted for at every node i of the riparian network. Simple mass balance gives the generalization of continuity (Equation 5.45):

$$\frac{d}{dt}\{S(t)\} + [C'']\{Q''(t)\} = \{I(t)\} \tag{5.140}$$

with the requirement of nonnegativity of both Q'' and S at all times:

$$\{Q''\} \geq \{0\} \tag{5.141}$$

$$\{S\} \geq \{0\} \tag{5.142}$$

A corresponding discrete-time form is

$$\{S^{k+1}\} - \{S^k\} + \Delta t [C'']\{Q''^k\} = \Delta t \{I^k\} \tag{5.143}$$

with superscript indicating time levels, separated by Δt.[5] Like the steady-state version, this continuity relation is not unique without the specification of the time-dependent decisions W and R, embedded in Q''. Continuity constrains the decisions,

---

[5] This is the simplest Euler discretization; there are many others, equivalent for small Δt.

as in the steady-state case. There is discretion for choosing the decisions W and R, which through the equality constraints affect the rest of the hydrology.

5.6.1 No Storage: The Quasistatic Case

To begin, revisit the single-farm hydroeconomy depicted earlier (Section 5.2.5), and introduce a wet and dry season $k = 1, 2$. All hydrological and economic variables will have two values; because there is no storage, the continuity equations will be uncoupled from each other, each season requiring its own static balance subject to the exogenous variation in resource availability. We will invent a third crop, X_a, to exercise this situation. Replicating the constitutive relation (Equation 5.50), we have

$$
\begin{bmatrix}
10 & 5 & 8 & -1 & 0 \\
1 & 1.5 & .5 & 0 & 0 \\
2 & 2 & 2 & 0 & 0 \\
e_c & e_b & e_a & -1 & 1
\end{bmatrix}
\begin{Bmatrix}
X_c^k \\
X_b^k \\
X_a^k \\
W_1^k \\
R_1^k
\end{Bmatrix}
\sim
\begin{Bmatrix}
\leq 0 \\
\leq L^k \\
\leq F^k \\
= 0
\end{Bmatrix}
\tag{5.144}
$$

coupled with the appropriate continuity relation for season k. If the economic programs in the two different seasons are completely separate, we simply have two separate programming problems: neither the networked hydrology nor the hydroeconomy exerts interseasonal constraints. Formally, these could be assembled:

$$
\begin{bmatrix}
10 & 5 & 8 & -1 & 0 & 0 & 0 & 0 & 0 & 0 \\
1 & 1.5 & .5 & 0 & 0 & 0 & 0 & 0 & 0 & 0 \\
2 & 2 & 2 & 0 & 0 & 0 & 0 & 0 & 0 & 0 \\
e_c & e_b & e_a & -1 & 1 & 0 & 0 & 0 & 0 & 0 \\
0 & 0 & 0 & 0 & 0 & 10 & 5 & 8 & -1 & 0 \\
0 & 0 & 0 & 0 & 0 & 1 & 1.5 & .5 & 0 & 0 \\
0 & 0 & 0 & 0 & 0 & 2 & 2 & 2 & 0 & 0 \\
0 & 0 & 0 & 0 & 0 & e_c & e_b & e_a & -1 & 1
\end{bmatrix}
\begin{Bmatrix}
X_c^1 \\
X_b^1 \\
X_a^1 \\
W_1^1 \\
R_1^1 \\
X_c^2 \\
X_b^2 \\
X_a^2 \\
W_1^2 \\
R_1^2
\end{Bmatrix}
\sim
\begin{Bmatrix}
\leq 0 \\
\leq L^1 \\
\leq F^1 \\
= 0 \\
\leq 0 \\
\leq L^2 \\
\leq F^2 \\
= 0
\end{Bmatrix}
\tag{5.145}
$$

Clearly, the two seasons are not coupled in any way. Each could be considered separately. Together, they amount to a quasistatic system, passing through a series of static hydroeconomic equilibria.

However, suppose that X_a requires a full year, while X_b and X_c may be different in the two seasons. This simple interseasonal coupling would require the added constraint

$$
X_a^1 = X_a^2
\tag{5.146}
$$

which would supplement the composite two-season economic relation (Equation 5.145) above. (We have interpreted the resource inputs for X_a as per season.) The added interseasonal constraint on economic activity (Equation 5.146) is the only interseasonal coupling. There are various condensations – for example, eliminating one variable X_a^2:

$$
\begin{bmatrix}
10 & 5 & 8 & -1 & 0 & 0 & 0 & 0 & 0 \\
1 & 1.5 & .5 & 0 & 0 & 0 & 0 & 0 & 0 \\
2 & 2 & 2 & 0 & 0 & 0 & 0 & 0 & 0 \\
e_c & e_b & e_a & -1 & 1 & 0 & 0 & 0 \\
\hline
0 & 0 & 8 & 0 & 0 & 10 & 5 & -1 & 0 \\
0 & 0 & .5 & 0 & 0 & 1 & 1.5 & 0 & 0 \\
0 & 0 & 2 & 0 & 0 & 2 & 2 & 0 & 0 \\
0 & 0 & e_a & 0 & 0 & e_c & e_b & -1 & 1
\end{bmatrix}
\begin{Bmatrix}
X_c^1 \\ X_b^1 \\ X_a^1 \\ W_1^1 \\ R_1^1 \\ \hline X_c^2 \\ X_b^2 \\ W_1^2 \\ R_1^2
\end{Bmatrix}
\sim
\begin{Bmatrix}
\leq 0 \\ \leq L^1 \\ \leq F^1 \\ = 0 \\ \hline \leq 0 \\ \leq L^2 \\ \leq F^2 \\ = 0
\end{Bmatrix}
\tag{5.147}
$$

Of course, this condensation does not change the coupling. Seasonal agricultural coefficients could be introduced that would not change the coupled structure either. Other forms of interseasonal crop requirements could be introduced – for example, a crop rotation requirement $X_b^1 = 5X_c^2$. The economic constraints are all that bind the seasons together; in terms of water management, we still have the quasistatic case.

More generally, interseasonal water storage will provide greater flexibility. The joint management of the hydrology and the economic activity requires the consideration of the fully dynamic system, as developed below.

5.6.2 The Reservoir

If we wish to alter the timing of naturally flows, then storage facilities will have to be present in the hydrological network. These are large, strategic works. They will have to be designed for certain performance criteria relative to the water uses made possible; then they will have to be owned and operated on fixed real estate in order to realize the design intentions. Significantly, they introduce the risk of catastrophic failure. These are two different contexts relative to the availability of information and the time scale relevant to decisions, and they require incorporation of some similar considerations at the fundamental level.

At the elementary level, a reservoir creates an impoundment or lake *a priori*, behind a structure. So in addition to the flow rates Q, W, etc., we need to add storage volumes S (and therefore depth of water). There are both opportunities – more flexibility in W, for instance – and liabilities, in the form of hazards, navigation impediments, ecological changes, land use effects, and the burden of operation. There are a host of other effects.

Consider a single isolated reservoir, with storage level $S(t)$; input and output flow rates $Q_i(t)$, $Q_o(t)$; and a withdrawal rate $W(t)$ as shown in Figure 5.5. The continuity equation is simple:

$$
\frac{\partial S}{\partial t} = Q_i - Q_o - W
\tag{5.148}
$$

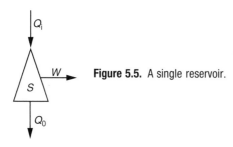

Figure 5.5. A single reservoir.

In discrete form, we will require a timing convention. Let k indicate a time interval of duration Δt. Then we have

Q^k: the *average* flow rate *during* time interval k.
S^k: the storage level *at the beginning* of time interval k.

A little care here with this simple convention eliminates many confusions later. The discrete form of continuity is then

$$S^{k+1} = S^k + \Delta t \left[Q_i^k - Q_o^k - W^k \right] \tag{5.149}$$

We retain the original distinction between flow rates and storage amounts; hence, Δt has compatible units. (The selection $\Delta t = 1$ ties the units to the time discretization, a convenience but again a potential source of confusion later.)

So far, we have treated the static and quasistatic cases. Next, we treat the climatological case, where a sequence of seasons repeats indefinitely, which is meaningful in a design-level search for sustainable operations. Ultimately, one must consider a facility as given and subject to significant departures from any climatological mean cycle. Catastrophic failure (whether hydrologic, economic, or ecological) due to extreme conditions is always possible; a sustainable situation must be robust in the face of such uncertainty about rare events.

Periodic Hydrology: Climatology

Suppose we have N seasons in a sequence that repeats indefinitely. Then the continuity relations would be

$$S^2 = S^1 + \Delta t \left[Q_i^1 - Q_o^1 - W^1 \right]$$
$$S^3 = S^2 + \Delta t \left[Q_i^2 - Q_o^2 - W^2 \right]$$
$$\vdots \ = \qquad\qquad \vdots$$
$$S^1 = S^N + \Delta t \left[Q_i^N - Q_o^N - W^N \right] \tag{5.150}$$

The last constraint embeds the assumption of periodicity. The N seasons generate N continuity constraints in $4N$ variables. To resolve the indeterminacy, we need to (a) provide constitutive relations for W and (b) connect the various flows to exogenous seasonal inputs in an appropriate network. It is assumed that all flow rates and storages are constrained to be nonegative.

This is the *climatological* view of hydrology: the same cycle repeating in every year, forever. Implied in this view is a repeating cycle of withdrawals and storages. The entire system is assumed to be in a periodic dynamic balance.

Storage-Yield

Suppose, for example, we have monthly climatological inflows as in Figure 5.6. A simple problem is to extract a constant monthly water supply W from this reservoir; to keep the storage within the size of the reservoir $\widehat{S}$; and to find a management regime that requires the minimum size reservoir. Because the minimum climatological inflow is 1.0, any $W \leq 1$ can be obtained with no storage whatever. For $W > 1$, storage capacity is needed. To the continuity equation, we add the constraint

$$S^k \leq \widehat{S} \tag{5.151}$$

and minimize $\widehat{S}$, given W. (Remember, the nonnegativity constraints apply to the hydrological variables.) Now the mean inflow is 5.42; so any W in excess of this is impossible on a sustained basis. So between the minimum Q_i and its mean, we can

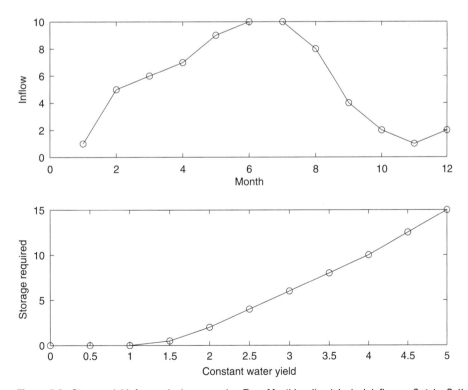

Figure 5.6. Storage-yield for a single reservoir. *Top:* Monthly climatological inflows, $Q_i \Delta t$. *Bottom:* Minimum reservoir capacity needed to guarantee $W \Delta t$ year-round.

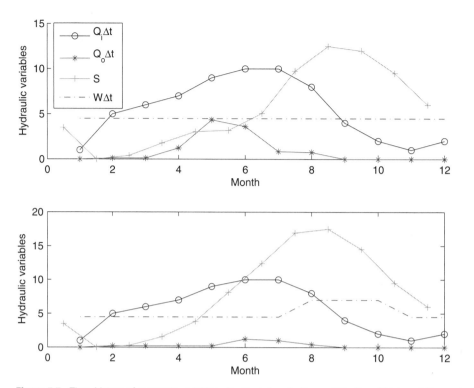

Figure 5.7. Time history of reservoir variables, for the inflow in Figure 5.6. *Top:* $W\Delta t = 4.5$, constant; *Bottom:* $W\Delta t = 4.5$, rising to $W\Delta t = 7$ during months 8, 9, and 10. According to convention, the flows are averages during a month, with S defined at the beginning of the month.

expect an increasing "storage-yield" relation, relating the smallest storage capacity $\widehat{S}$ necessary to guarantee a given constant W. This is plotted in Figure 5.6. Implied is a monthly management regime for S and Q_o. An example of this appears in Figure 5.7 for $W\Delta t = 4.5$. These optimizations are achieved with the Excel Solver, program **StorageYield.xls.**

This framework admits many complications. Suppose, for example, that water demand is higher during a peak season. This will require more storage capacity and a different management regime. The case $W\Delta t = 4.5$ normally, rising to 7 during three months, is shown in the lower panel of Figure 5.7. In the reference case (top panel), there is significant outflow (e.g., during month 5). In the "peaking" case, this is instead stored in anticipation of the later peak withdrawal. The required storage is higher: $\widehat{S} = 17.5$, versus 12.5 in the reference case.

A further perturbation would be to require a partly empty reservior during periods of high flood potential. If the flood protection is posed as a necessary dry volume ΔS be kept empty during a certain season, then the constraint $S^k \leq \widehat{S} - \Delta S$ must be enforced in that season.

Finally, note that solutions with significant W use much of the inflow leaving the optimal output flow Q_o very low or zero in several months. One can constrain this to be nonzero; that would require higher storage capacity and/or an adjusted storage regime. Constraints could also be posed to limit the high and low values of S. These adjustments can be explored within the framework of program **StorageYield.xls**.

Water Supply and Power

Again with a single reservoir, consider the case where hydropower equipment is built into the facility. In the analysis in Section 5.2.7, the power generation rate $\dot{E}$ will be the product of pressure and flow rate, the pressure a monotone function of the storage via the storage head relation $H = H(S)$:

$$\dot{E} = 9.81\epsilon Q_o H(S) \tag{5.152}$$

where we have assumed that all of the downstream flow Q_o passes through the hydropower machinery. (This will require a maximum possible Q_o, reflecting the design particulars.) The annual energy generation will be the time integral of $\dot{E}$; in discrete form,

$$E = 9.81\epsilon\Delta t \sum_k Q_o^k[H(S^k) + H(S^{k+1})]/2 \tag{5.153}$$

where the superscript indicates time, summed over one climatological year.[6]

Assuming a known or hypothesized facility size, there will be constraints on the maximum and minimum S possible (failure due to overtopping; minimum reservoir level) and on the maximum and minimum possible outflows based on the machinery design. These same variables, S and Q_o, can have additional ecological, economic, or political constraints, as well as consideration of the possibility of extreme events. These all lead to the same forms:

$$\underline{S} \leq S^k \leq \overline{S} \tag{5.154}$$

$$\underline{Q_o} \leq Q_o^k \leq \overline{Q_o} \tag{5.155}$$

And, of course, nonegativity is required in the usual way. And so we arrive at a multipurpose form of the reservoir problem: Given a climatological inflow pattern $Q_i(t)$, a desired $W(t)$ pattern, and a given hydrostorage facility, manage the storage history so as to maximize the power production.

It is clear that water supply and energy production are at odds with each other here. All W is flow that does not generate any energy, bypassing the machinery. Flow goes either to W or to $\dot{E}$, but not to both. It is also clear that power generation itself would prefer a constant large storage (high pressure), while the water supply interest prefers considerable manipulation of storage level to smooth out inflow variations.

The trade-off between yearly water supply and yearly hydropower production is illustrated in Figure 5.8. In this case, the inflow climatology from the previous example is used, with a fixed facility ($2 \leq S \leq 10$; $1 \leq Q_o\Delta t \leq 10$). For each point, the value of

---

[6] This is a discrete approximation to the integral, using a version of the trapezoidal rule. Other options are clearly possible for discretizing the integral.

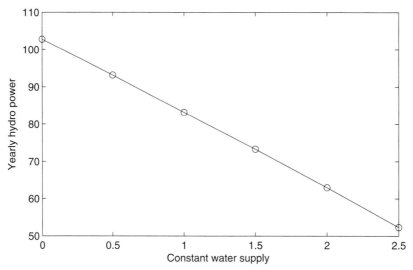

Figure 5.8. Water supply – hydropower trade-off. The storage–head relationship is quadratic: $H \sim S^{0.5}$.

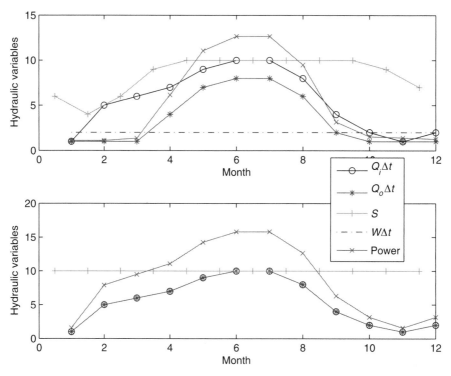

Figure 5.9. Single-reservoir hydropower. Climatological pattern of storage and flows, for $W\Delta t = 2$ (*top*) and $= 0$ (*bottom*). Power units are scaled to fit the graph conveniently. In the bottom panel, $Q_i = Q_0$ and $W = 0$.

W is fixed (a constant) and the storage regime is manipulated to produce maximum annual energy. The correpsonding annual storage cycle is shown in Figure 5.9, for a high value of water supply, $W\Delta t = 2$, and the opposite extreme of no water supply. The latter is intuitive: Because the raw inflow is accommodated by the machinery, there is no need for storage; so the reservoir is kept completely full for maximum pressure power: $Q_i = Q_o$. The high water supply case is significantly different, with very dynamic storage regulation.

In this problem, there is no preferance for energy in any one season; all seasons are equivalent, so the result has dramatic power peaking during months 6 and 7 and constant W. Clearly, other desirable patterns can be expressed via the constraints.

There would be a different set of graphs for each possible reservoir/hydro-combination. The program **Hydro1_impl.xls** optimizes this case.

5.6.3 Two Reservoirs

Suppose we have multiple reservoirs in a network. Naturally, these will require joint operation, as upstream management activity affects the downstream options and constraints.

Consider the simple two-reservoir system in Figure 5.10. There is installed hydro-electric capacity at both; neither has any water withdrawal. The continuity relations will be

$$\frac{dS_1}{dt} = -Q_1 + Q_{i1} \tag{5.156}$$

$$\frac{dS_2}{dt} = -Q_2 + Q_1 + Q_{i2} \tag{5.157}$$

Reservoir 1 will process all of Q_{i1} through its turbines; Reservoir 2 will process all of $Q_{i1} + Q_{i2}$. The joint system will operate in an annually repeating cycle, with climatological inputs.

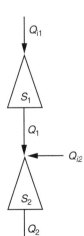

Figure 5.10. Two reservoirs in series.

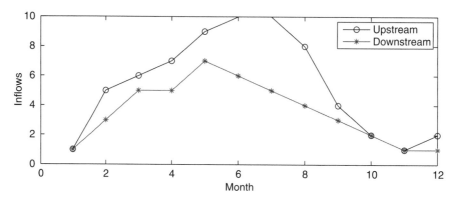

Figure 5.11. Climatological flows into the two-reservoir system: Q_{i1} (upstream) and Q_{i2} (downstream).

For this example, the upstream dam is identical to that in the previous example, $2 \leq S_1 \leq 10$ and $1 \leq Q_1 \leq 10$; and the input Q_{i1} is unchanged. The downstream dam is larger, with limits $4 \leq S_2 \leq 20$ and $2 \leq Q_2 \leq 10$ and $H \sim S^{0.7}$. Both have the same electrical efficiency. Climatological inputs appear in Figure 5.11. The downstream tributary is smaller, but it is joined by the upstream reservior outflow and therefore the larger reservoir/turbine combination downstream.

The maximization proceeds as above, this time maximizing the joint annual production of energy under periodic operation and no water withdrawal. The solutions appear in Figure 5.12. A salient feature is the full reservoir downstream during much of the cycle, reflecting the unilateral focus on energy. As there is no water supply withdrawal, it is interesting to compare the upstream operation with the one-dam optimization of the same case (Figure 5.9, bottom). Clearly, the upstream releases are timed with the downstream generation and constraints in mind.

Program **Hydro2_impl.xls** treats this case.

There are many variants combining ideas embedded in these simple examples. Common are the specifics of the networked hydrology; the storage and flow variables; the valuable nature of the water; and the attempt to match natural variability in inflow, to human needs in terms of predictability and smoothness; and the necessary satisfaction of constraints of many types on flows and storage levels.

5.6.4 Simulation: Synthetic Streamflow

In simulation studies, one can represent many complexities in increasing realism, including the selection of design parameters, the operation of systems, the forecasting of inputs, the effects of different operating rules on outcomes, and the value added by additional sophistication in management tools or observation networks. One of the critical issues is foundational: the inputs of water to any system.

Real inflow sequences never repeat precisely. Therefore, in studies of design and operation of water resource systems, we are always abstracting the situation to the case of idealized flow sequences and their management. A procedure that uses only

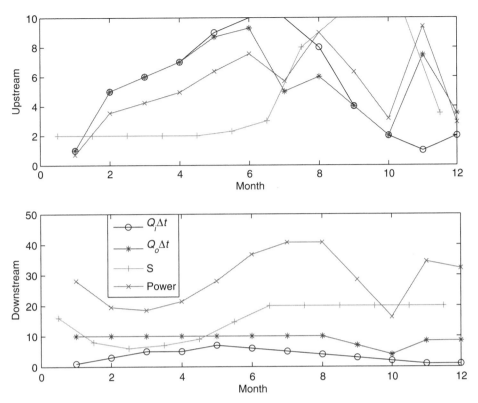

Figure 5.12. Energy-maximizing solution for the two-reservoir example. (Power units are scaled to fit the display.)

an observed time series is deficient in three ways: It is limited by the length of the time series, which in many cases may be shorter than the design life under consideration; it enshrines vagaries of the data, including measurement error and simple gaps in the record; and finally, it will *never* occur again.

Instead, we hope to synthesize time series that are believed to be statistical replicas of data and to subject any analysis to several such time series, hoping to find robust features, operating rules, and freedom from extreme or unusual occurrences. The greater the fidelity, statisically, of synthetic series to truth, the better will be the outcomes. This means, of course, a dependency of the entire procudure on (a) having enough data to extract its important statistical properties; and (b) using a sufficiently large "ensemble" of time series to generate a representative ensemble of outcomes.

It is reasonable in hydrology to find a climatological mean tendency. But about this, one can expect disturbances that can be understood only in a probabilistic sense, not mechanistically. A simple version of this might be

$$Q(t) = \overline{Q} + \widehat{Q}(t) \tag{5.158}$$

where we have separated the constant and time-variable parts. It is common to expect to observe the constant part through averaging and to observe only the statistics of the time-variable part. This would be commonly realized in a time series:

$$Q^k = \mu + \mathcal{R}^k \qquad (5.159)$$

where μ is the observed mean[7] flow,

$$E[Q] = \mu \qquad (5.160)$$

and $\mathcal{R}^k$ is a random process with no memory, zero mean, and statistical properties representative of flow departures from the mean. There are infinitely many such time series, all equivalent in that they share the same origin and statistical properties. In the Appendix, we refer to a few common distributions; the Gaussian, or normal, distribution is common and useful to fix ideas.

Autocorrelation

The process in Equation 5.159 has zero memory – that is, the probability of something happening next is independent of what has already happened. But the variability $\widehat{Q}(t)$ may well be simply masking natural processes that we simply do not understand or otherwise cannot predict. Therefore, we can expect dynamic processes in its origin, resulting in characteristic time scales of interdependence in $\mathcal{R}$ and therefore Q. The simplest generator is the first-order Markov model,

$$Q^k = \mu + \rho(Q^{k-1} - \mu) + \mathcal{R}^k \qquad (5.161)$$

where the parameter ρ is the autocorrelation coefficient. If the expected value of Q^k is always μ in the case of no memory ($\rho = 0$, Equation 5.159), here the expected value is conditional, depending on the most recent history:

$$E[Q^k|Q^{k-1}] = \mu + \rho(Q^{k-1} - \mu) \qquad (5.162)$$

$$= \rho Q^{k-1} + (1 - \rho)\mu \qquad (5.163)$$

Essentially, Q^k "remembers" the previous departure from the long-term mean, ρ is the strength of the memory. The long-run, or unconditional, mean remains μ as above,

$$E[Q] = \mu \qquad (5.164)$$

irrespective of ρ; but it requires a longer time series to manifest, due to the memory. The second moments are sensitive to ρ:

$$E[(Q - \mu)^2] = \frac{E[\mathcal{R}^2]}{1 - \rho^2} \qquad (5.165)$$

Hence, if σ^2 is the variance of Q, then the variance of $\mathcal{R}$ is reduced:

$$\sigma_{\mathcal{R}} = \sigma_Q \sqrt{1 - \rho^2} \qquad (5.166)$$

---

[7] We employ the expectation (statistical average) operator E.

Finally, the covariances of Q are

$$E[(Q^k - \mu)(Q^{k-1} - \mu)] = \rho\,\sigma^2 \tag{5.167}$$

$$E[(Q^k - \mu)(Q^{k-n} - \mu)] = \rho^n\sigma^2 \tag{5.168}$$

Hence, given a time series Q^k, the parameters of the lag-1 Markov model may be estimated in the sequence of moment calculations:

$$\mu = E[Q] \tag{5.169}$$

$$\sigma^2 = E[(Q - \mu)^2] \tag{5.170}$$

$$\rho = E[(Q^k - \mu)(Q^{k-1} - \mu)]/\sigma^2 \tag{5.171}$$

and the streamflow generator is

$$Q^k = \mu + \rho(Q^{k-1} - \mu) + \sigma\sqrt{1 - \rho^2}\,\mathcal{R}^k(0,1) \tag{5.172}$$

where $\mathcal{R}^k(0,1)$ is, as before, random, uncorrelated, and zero mean; the proviso is added here that it has unit variance. (The two arguments of $\mathcal{R}$ indicate its mean and variance.)

Climatological Mean

A simple generalization of the above involves the assertion that μ itself is time-varying – for example, periodic climatological mean μ^k. Earlier, we assumed that this was cyclic and subject to no randomness. Here we assume that observations can be demeaned by month or season, and the departure from that mean is random and autocorrelated. Thus,

$$Q^k = \mu^k + \rho(Q^{k-1} - \mu^{k-1}) + \sigma\sqrt{1 - \rho^2}\,\mathcal{R}^k(0,1) \tag{5.173}$$

The procedure described above follows; instead of first estimating μ, it is estimated by month or season, μ^k.

Example

In Figure 5.13, the monthly observed flowrate of the St. John River is plotted versus time for 11 years. The mean climatological tendency is evident, as is the variance about it. There is a major peak in the spring and a secondary peak in the fall. The lower panel in Figure 5.13 shows the climatological mean cycle: each datum in this panel is the mean of 11 representations. Departures from this climatoligical cycle appear in Figure 5.14. Visual inspection reveals significant variation, mild memory, and both wet and dry periods – mild but sustained.

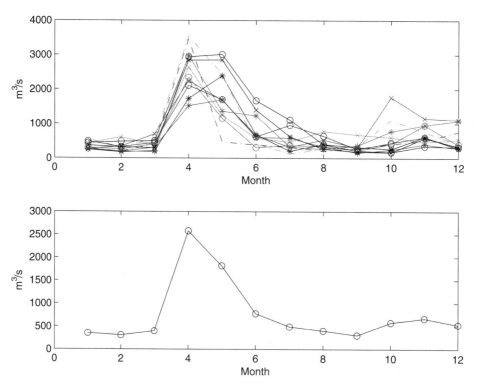

Figure 5.13. St. John River: 11-year climatology. *Top*: individual yearly records; *Bottom*: climatological mean cycle.

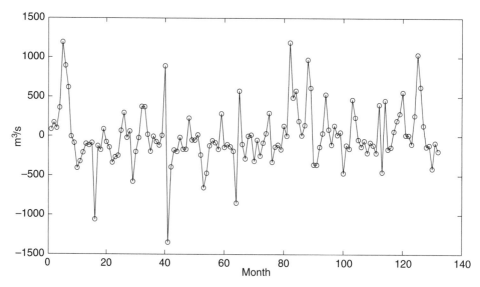

Figure 5.14. St. John River: arithmetic departures from climatology.

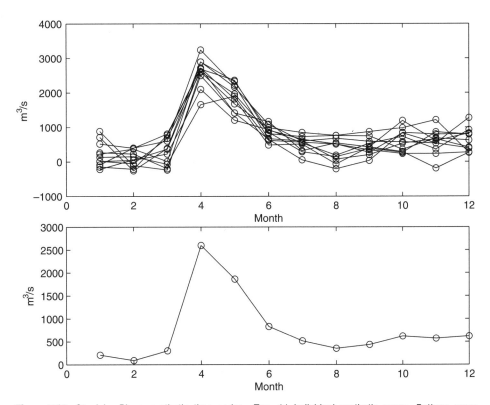

Figure 5.15. St. John River: synthetic time series. *Top*: 11 individual synthetic years; *Bottom*: mean synthetic "climatology."

These departures from climatology were analyzed as described above, to produce the values $\sigma = 373$, $\rho = .29$. These values were then used with a Gaussian $\mathcal{R}$ to produce 10 years of synthetic streamflow (Figure 5.15). The climatological mean of this synthetic decade appears in the lower panel of Figure 5.15. The correspondence is reasonable. The large peak is more coherent and more precisely timed than that in the data. The dry period generated between the peaks is too variable. In addition, the synthetic mean is "bumpy" in this period. Significantly, the synthetic hydrology produces negative Q occasionally, there being nothing to prevent this.

Naturally, each new calculation produces a unique new sequence, statistically the same. The rough spots indentified above show a need for a larger ensemble.

Data for this example and the analysis itself can be found in the program **StJohnStatistics.xls**. Matlab implementations are **stjdata.m**, **stjeps.m**, and **stjsyn.m**.

Logarithmic Transformation

It is common in hydrology to apply various mathematical transformations to the primitive variable Q (streamflow rate), such that it cannot be negative by assumption. The lognormal transformation is an example: If the logarithm of Q is normally distributed, then Q itself, being the exponential of a real number, is always positive

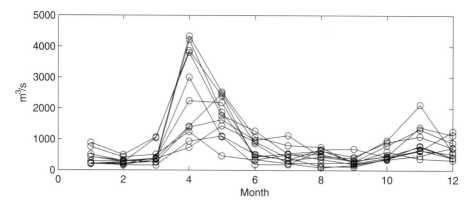

Figure 5.16. St. John River: synthetic hydrology with a lognormal, lag-1 Markov generator.

by hypothesis. So a common adaptation would be to transform the data to Y:

$$Y = \ln(Q - \tau) \tag{5.174}$$

$$Q = \tau + e^Y \tag{5.175}$$

and find the statistical properties of Y. The assumption that Y is Gaussian amounts to the common lognormal distribution of Q. Relations among the moments of Y and Q appear in the Appendix. (τ is a new parameter, the absolute minimum possible value of Q.)

Repetition of the above example with the lognormal assumption leads to the generator

$$Y^k = \mu_y^k + \rho_y(Y^{k-1} - \mu_y^{k-1}) + \sigma_y\sqrt{1 - \rho_y^2}\,\mathcal{R}^k(0, 1) \tag{5.176}$$

$$Q^k = \tau + e^{Y^k} \tag{5.177}$$

This would be a lognormal, lag-1 Markov synthetic streamflow generator. As suggested by the nomenclature, the parameters $(\mu_y, \rho_y, \sigma_y)$ are estimated from the transformed data Y; the fourth parameter τ is estimated as the minimum possible value of Q; and $\mathcal{R}$ is assumed normal.

Figure 5.16 illustrates this synthesis, for $\tau = 0$, $\sigma_y = .456$, and $\rho_y = .445$. The negative flows are gone. There is a greater tendency toward very high flows at the spring peak (not evident in the ensemble shown). Programs **StJohnStatisticsLN.xls** and **stjlnsyn.m** illustrate this calculation.

Pioneering work in this area is available in Fiering [24, 25]. An excellent presentation in greater depth can be found in Loucks et al. [53].

5.7   CASE STUDY: THE WHEELOCK/KEMENY BASIN

Consider the Wheelock/Kemeny river system as shown in Figure 5.17. There are six hydrological seasons. Two impoundments have been created at the Rockefeller and

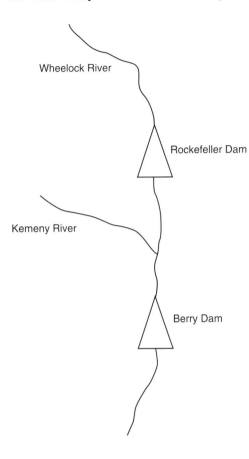

Wheelock River

Rockefeller Dam

Kemeny River

Figure 5.17. Wheelock–Kemeny Basin.

Berry Dam

Table 5.7. Data for the Wheelock/Kemeny Basin example: inflows and hydro release limits.

	Upstream			Downstream		
	$Q_i \Delta t$ (Wheelock)	$Q_o \Delta t$ (Rockefeller)		Q_i (Kemeney)	Q_o (Berry)	
Time		min	max		min	max
1	60	20	90	50	30	140
2	50	20	90	40	30	140
3	40	30	90	30	40	140
4	80	30	90	60	40	140
5	100	30	90	75	40	140
6	120	20	90	90	30	140

Berry sites; each has hydropower equipment installed. The general objective is to maximize annual energy production.

The storage–head relations are

$$\frac{H}{\overline{H}} = \left(\frac{S}{\overline{S}}\right)^{\alpha} \qquad (5.178)$$

Table 5.8. Data for the Wheelock/Kemeny Basin example: dam and turbine parameters. α is the storage head parameter; ϵ is the electrical efficiency.

	Upstream S_1 (Rockefeller)	Downstream S_2 (Berry)
α	0.64	0.55
ϵ	0.8	0.85
$\bar{S}$	150	400
$\underline{H}$	30 m	60 m
$\overline{H}$	70 m	90 m

Data are given in Tables 5.7 and 5.8. Several example problems employ this example. The case is adapted from one in Loucks et al. (1981).

5.8 RECAP

Early work in water resources was recorded in the seminal works of Fiering [25, 24], Loucks et al. [52], Hufschmidt [44], Marglin [61], Major [57], and others.

There are many excellent contemporary works that build on these early beginnings. Shiklomanov and Rodda [81] offer comprehensive assessment of global hydrology. Loucks et al. [53] present a comprehensive coverage of water systems analysis. Sustainability of water resource systems is addressed by Loucks and Gladwell [51, 2]. Bogardi and Kundzewicz [4] examine risk assessment for water resource systems; Griffin [35] offers a good treatment of water resource economics. Gleick provides a regular biennial assessment of water issues [31].

The early work was inspired by the international development imperative; that remains as a catalyst. Contemporary attention to the legal and institutional frameworks within which one operates is available in Salman and Bradlow [76]. Water is among the millenium development goals (Gleick [30]) and part of the comprehensive United Nations World Water Program [95]. Access to water has been recognized as a universal human right [77, 28, 29], reflecting its importance to human habitat, food production, and public health. This is the ultimate assertion of value.

5.9 PROGRAMS

WaterLand.xls. Linear-programming (LP) formulation for static allocation of three resources among 2 activities.

TwoFarms.xls. LP formulation for static allocation of water and other resources among two riparians (Section 5.2.5).

WentworthN.xls. LP formulation for the Wentworth Basin, cases $N = 1-6$ (Section 5.3).

StorageYield.xls. LP formulation for the storage-yield examples shown in the text.

Hydro1.xls. NLP optimization of hydropower production at a given facility, subject to known inflow pattern and constant water supply withdrawal.

Hydro1_impl.xls. The same as **Hydro1.xls**, except the integration of annual power is more precise, employing the trapezoidal rule.

Hydro2.xls. NLP optimization of a two-dam system, as in Section 5.6.3.

Hydro2_impl.xls. Same as **Hydro1.xls**, except the integration of annual poser is more precise, employing the trapezoidal rule.

StJohnStatistics.xls. Lag-1 Markov generator and the St. John discharge data.

StJohnStatisticsLN.xls. Lognormal form of the St. John analysis.

stjdata.m. The data for the St. John example.

stjeps.m. The St. John departures from its climatological mean.

stjsyn.m. The synthetic generator for the St. John example.

stjlnsyn.m. The logarithmic synthetic generator for the St. John example.

5.10 PROBLEMS

1 Develop the shadow prices for the four optimal solutions A, B1, B2, and C depicted in Figure 5.2 and Table 5.4. Make a table equivalent to Table 5.3 for this case and compare it with the case when F is huge.

2 Define a third form of continuity equation in which W, E, and R are all defined at each point of withdrawal. Do this as an extension of the existing forms 5.41 and 5.44 for the example in Figure 5.3. Show an accompanying adjustment to the two-farm hydroeconomy relation (Equation 5.52), which uses E_i instead of R_i.

3 Change Equations 5.57 and 5.58 such that the fertilizer used on the upstream farm is made in the downstream city and is labor intensive. One unit of fertilizer requires 30 units of population.

4 Revisit the Wentworth Basin (Section 5.3). Solve Case 1 using linear programming. Then solve these problems:
  (a) The downstream nation wishes to sell back its treaty rights. What price would the upstream nation be willing to pay per unit of water?
  (b) An expensive interbasin transfer of water is suggested from an adjacent basin, supplementing I_a. What cost of this project, per unit of water, is justified?
  (c) An endangered species has been discovered just downstream of both farms diversions (the reaches between W_1 and R_1 and between W_2 and R_2). It is proposed to double the minimum flow requirements there. What is the cost of this regulation?

5 Revisit the Wentworth Basin (Section 5.3). Solve Case 4 using LP for several values of the ethanol subsidy S, and plot the optimal withdrawal and employment levels (total and ethanol-related) versus S.

6   Revisit the Wentworth Basin (Section 5.3). Solve Case 5 using LP for several different values of power generated at Q_8. At what point does power generation cease to dominate all other uses? Plot total employment (labor) versus power production for these cases.

7   Revisit the Wentworth Basin (Section 5.3), Case 5 with no ethanol subsidy. A new law has been enacted that makes export of power illegal. What is the basin program now? How much power can be generated at the extant site without affecting the rest of the basin program?

8   Revisit the Wentworth Basin (Section 5.3). Compare the optimal solutions for Case 3 (maximize money) with Case 6 (maximize employment). The text already compares the optimal programs; make a comparison of the constraints – that is, which ones bind, which are slack in each case?

9   Revisit the Wentworth Basin (Section 5.3). Compare the optimal programs for Case 6 (maximize employment) under two conditions: (a) money is constrained to exceed 0.20; and (b) money is required to be nonnegative. Note and explain any differences with Case 6 recorded in the text.

10   Revisit the Wentworth Basin, using Case 3 results as the starting point. A fish ladder is required (see Figure 5.18). The flow in this ladder is W_6. The requirement is $W_6 \geq 0.18$. What is the cost of this requirement in terms of lost production in the basin?

11   Revisit the Wentworth Basin, using Case 3 results as the starting point. A canal with locks is proposed (Figure 5.18). (The fish ladder from Problem 10 has been abandoned.) The flow through the canal is W_7. If built, the canal capital cost would be $K = \$400$ per year, and it would accommodate up to 800 vessels per year. Each vessel requires nonconsumptive flow through the canal and locks of .002 units of water. Fee per vessel is to be negotiated; treat it as a parameter.
   (a) Describe the extra constraints that need to be added to Case 3 to represent this.
   (b) What fee (per vessel) will justify building the canal? What fee will guarantee its full utilization?
   (c) What is the effect of the fees in (b) on the rest of the basin economy?

12   Examine the Freedman Basin in Figure 5.19. The agricultural technology is given in the table below:

	Cotton	Barley	Wool
Water	10	5	0
Land	1	1.5	0
Labor	20	20	5
Sheep	0	0	100
Value	5	10	5

The population of the cities is 50 (C1) and 100 (C2). People unemployed in agriculture are assumed to be shepherds. There is a policy of full employment. The pasture for

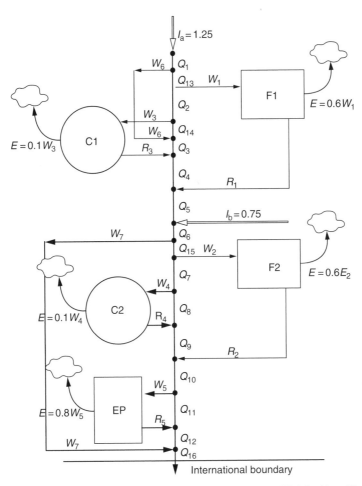

Figure 5.18. Wentworth Basin network details with fishladder (W_6) (Problem 10) and navigation canal (W_7) (Problem 11) added.

sheep is not counted in the arable land at the two farms; there is enough to support 1,000 sheep but not more.

The city requirements: Withdrawals must exceed .005 per person; the population is fixed, and only City 1 residents can work on Farm 1 and likewise for City 2/Farm 2.

Withdrawal costs: zero in all cases except W_5 and W_6. These require pumping over a mountain range, with an energy cost of $3.1 per unit pumped. In addition, if W_5 is used at all, there is a fixed cost of $4 that must be invested in land acquisition along the pipeline route. (Hint: use an integer variable.)

Flow requirements: All flows in natural channels (not the withdrawal pipes) must exceed 0.3

Downstream: A treaty requires 50% of the total available water (that is, 0.9 units) to be delivered.

   (a) Formulate this as a network optimization problem. Be clear about constraints, objectives, and decision variables.

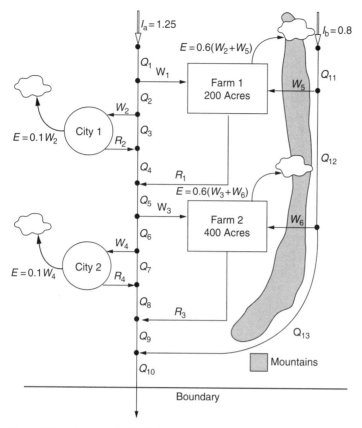

Figure 5.19. Freedman Basin, Problem 12

.

(b) Solve. What is the cropping pattern, the number of sheep and shepherds, and the withdrawals?

(c) Which constraints are slack?

(d) Suppose a bus system is proposed between the two cities. Now workers may move freely between the two farms. What is the value of this transit system?

13 A development agency will allocate $10 Billion among four types of public investment: defense (D), civil works (C), education (E), and land reform (L). There are three economic sectors that will benefit: the farm sector (F), the urban/industrial sector (U), and the rural nonmarket sector (N). The total benefits for each sector stemming from a unit of public investment are as follows:

$$Z_F = 1D + 4C + 3E - 10L \tag{5.179}$$

$$Z_U = 1D + 6C + 6E + 0L \tag{5.180}$$

$$Z_N = -1D + 2C + 8E + 10L \tag{5.181}$$

There are two objectives: (a) Maximize the total benefit; and (b) Keep inequity low: the range R among the three benefits must be below 20% of the mean of the three.

One unit of R above the goal is equivalent to 5 units of total benefit. *Formulate this as a linear goal programming problem.*

14 The operators of a ski resort have three basic decisions to make: the cost C of a season's pass; the amount S to invest in snowmaking; and the amount F to invest in lodge improvements. There are three metrics of success: P = public participation (skier days); L = financial loss (\$K); and D = the duration (days) of the tourist season. These are related linearly to the decisions as follows:

$$P = 15{,}000 - 100C + 50S + 40F \tag{5.182}$$

$$L = 200 - 5C + S + F \tag{5.183}$$

$$D = 60 + 10S \tag{5.184}$$

Management wants to satisfy the following:
  (a) Strictly avoid a loss exceeding \$400K.
  (b) Stay open at least 80 days; the penalty for early closure is \$80K/day.
  (c) Achieve participation of at least 30,000; the penalty for lower participation is \$50 skier-day.
Formulate this as a linear goal programming problem.

15 A development agency has at its disposal limited amounts of money (\$300K), equipment (20), and experts (10). Its mission is to increase employment in both agriculture and manufacturing, and it will allocate its resources among small projects in either sector. Let x_a and x_m be the number of such projects, and assume that fractional projects are meaningful. Each agricultural project requires \$60K and one expert, and generates 200 jobs. Each manufacturing project requires \$20K, two experts, and five machines and would generate 300 jobs.

The Goals are to generate at least 500 jobs in agriculture and 900 jobs in manufacturing. New jobs in excess of these targets are valued twice as highly in agriculture as in manufacturing. Unused money is also valuable, and \$10K unused is equivalent to one new agricultural job. Spare equipment and spare experts have no value.
Formulate this as a linear goal programming problem.

16 Revisit the storage-yield calculations shown in Section 5.6.2, with the "tourist season" augmentation to the required water supply. Add the requirement of flood protection during months 3–9, as described in the text. Discuss the solution properties and plot the required storage capacity versus the amount of flood protection provided.

17 Revisit the basic storage-yield calculation (Section 5.6.2) for $W = 3$ year-round. Implement a year-round minimum-flow requirement downstream of the dam: $Q_o \geq$ the minimum flow observed in the input hydrology. What is the effect on $\widehat{S}$? Explain the time-history of storage and release by comparison with the base case of no minimum flow requirement.

18 Redo the trade-off between energy and water supply (Figure 5.8). Assume hydropower is required to exceed a constant monthly baseline. Plot the baseline power value versus the constant water supply provided. Discuss relative to the case in the text.

19 Build a random-number generator for the Gaussian distribution, using the uniform deviate generator (See the Appendix). Test it as follows: Generate an ensemble of N Gaussian deviates; calculate the mean and the standard deviation of the ensemble; plot the mean and standard deviation versus N.

20 Consider the streamflow sequence $Q = [5, 6, 7, 6, 5, 4, 2, 4, 3, 1, 1, 5, 4, 2, 2, 1]$.

   (a) Fit a lag-1 Markov model to these data. Report the values of μ, σ, and ρ.

   (b) Generate an ensemble of 10 synthetic flow sequences of the same length. Plot the ensemble and the original data; compare and discuss.

   (c) For each synthetic sequence generated, estimate the values of μ, σ, and ρ. There will be 10 such estimates of each parameter. What are the mean and variance of each parameter estimate?

   (d) Synthesize a single long flow sequence of length N. From this sequence, compute and plot an estimate of the three parameters μ, σ, and ρ. Plot the results versus N. What do you expect, and does this experiment confirm that?

21 The Wheelock/Kemeny Basin case study: Consider the Rockefeller Dam in isolation. Find the optimal annual hydraulic cycle and energy production. Confirm that the annual cycle satisfies continuity and violates no constraints.

22 The Wheelock/Kemeny Basin case study: Consider both facilities. Use the result of Problem 21 for managing Rockefeller, and invent three different ad-hoc cycles for Berry that are credible. Evaluate the performance of each by reporting (1) the annual energy production and (2) the failure rate for any of the constraints that are violated.

23 The Wheelock/Kemeny Basin case study: Find the optimal operating policy for both facilities considered jointly. Report the energy production and the annual cycle (both hydraulic and electric). Has the operation of the Rockefeller Dam changed from that found in Problem 21?

24 *The Wheelock/Kemeny Basin case study:* A fish ladder is proposed at Berry. It will require the continuous release of $Q_f \Delta t = 30$, which will bypass the turbines, in addition to the minimum release requirements shown. What is the impact on energy production? If none, how high can the flow $Q_f \Delta t$ be before an impact is felt?

25 Examine the Salmon River and Dam shown in Figure 5.20. The river has seasonally variable streamflow. Two crops can be grown; they have seasonal water needs per ton of crop. Additionally, there are seasonally varying demands for flow through the fish ladder, and the value of power produced at the dam is seasonal. All these data are as follows:

Season	I	Ladder	Water/Ton Wheat	Corn	Power Value
1	10	0	5	6	30
2	25	10	0	2	20
3	20	10	0	0	20
4	5	0	7	5	30

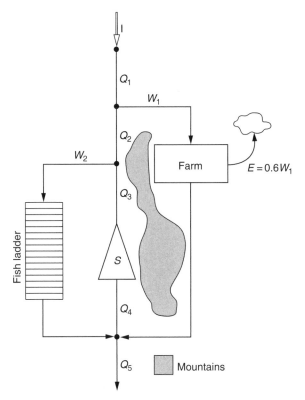

Figure 5.20. Salmon River and Dam, Problem 25.

Land available is 1,000 units. Other agricultural data are seasonally invariant, as shown below:

	Wheat	Corn
Land/Ton	2	1.25
Crop Value/Ton	20	30

The dam has these characteristics: $H = S^{0.33}$; $10 \leq S \leq 50$; $5 \leq Q_4 \leq 20$; and $\dot{E} = .9HQ_4$.

    (a) Formulate this as a nonlinear programming problem. Show explicitly the decision variables, the constraints, and the objective.

    (b) Solve. What is the seasonal pattern of power production, wheat, corn, and unused land?

    (c) Suppose the fish ladder were dismantled. What is its value, in terms of power, corn, and wheat?

26 Consider the Whitehall Locks and Canal (Figure 5.21). There is a dam, a farm, and a navigation canal. There are two seasons, winter and summer, with $Q_{1w} = 25$ and $Q_{1s} = 10$. The seasons cycle indefinitely. All flow-related variables will have two seasonal values. The farm has two crops: oats in the summer and wheat in the winter. In additon, it sells wool, for which it keeps a year-round herd of sheep, which needs

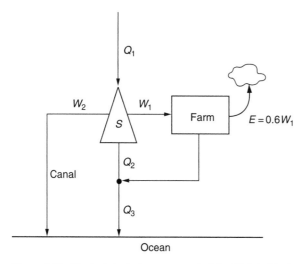

Figure 5.21. The hydrological arrangement at the Whitehall locks, Problem 26.

land and water. There are 50 hectares of land available. The crop requirements and values are:

- oats, per hectare: 5 units of land, 1 unit of water, $30 per hectare
- wheat, per hectare: 2 units of land, 3 units of water, $35 per hectare
- sheep, per head: .2 units of land, .1 unit of water, $20 of wool per head, per season

At the dam, storage must be kept within 20 and 40 units of water; maximum release is 20 units; there is no minimum release. The storage-head relation is quadratic. The power is worth $5 per kilowatt in the wet season, $7 in the dry season. Downstrean, the natural flow to the ocean must exceed 3.0 at all times. It is desired to optimize this basin. Navigation value depends on the canal flow, at $78 per unit of flow.

Your job is to formulate this as a linear-programming problem. Follow these steps:

    (a) List all the *decision variables* you need. Which ones cannot be negative?

    (b) Write out the constraints on the hydrology.

    (c) Write out the constraints on the farm.

    (d) Write out the constraints on the dam storage and release.

    (e) Write out the constraints on the dam operation (storage and release requirements).

    (f) Write out the economic objective that is to be maximized.

27 An isthmus separates two coastal seas A and B (Figure 5.22). It is very rainy, and there is a dense rain forest draining to a central Lake Ganut, well above sea level. A navigation canal crosses the isthmus, with a complex series of locks. The lake flows out partly through the locks, enabling navigation. The navigation flow Q_N is split as shown in the figure, and every vessel must go either from A to B or vice versa via Lake Ganut. Excess water exits directly to the sea either through a hydropower facility Q_H or over a spillway Q_S. See the illustration.

Inflow to the lake has three seasons: $Q_{in} = (1, 8, 6)$. It is periodic.

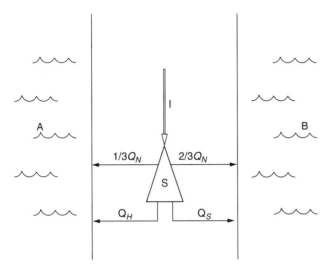

Figure 5.22. The high Lake Ganut and canals to the sea, Problem 27a, b.

Canal use is in high demand year-round. The maximum navigation flow in any season is 3 units. Navigation use is valued at P_N per unit of flow. The year-round maximum accounts for 9 units of flow at most.

The storage-head relation in the lake is $h = S^{0.6}$. The storage must always be between $S_{\min}$ and $S_{\max}$.

Hydropower flow Q_H is limited to 1 unit in any season. The spillway flow Q_S cannot occur unless the lake is full; further, it must not exceed 4 units in any season. The only client for hydropower is the servicing of the vessels in the locks. Hence, power must be proportional to the navigation flow in any season: Power $= .2 * Q_N$.

The objective is to maximize the annual navigation income:

  (a) Formulate (do not solve) this as a nonlinear programming problem.
  (b) International contracts require that the three seasonal navigation flows not vary by more than 30% of their average. Modify problem (a) to reflect this.
  (c) It is proposed to add an upstream dam (Figure 5.23) in order to provide better interannual water management. Only one design is feasible, to provide storage capacity $\overline{S}_u$ at expense K_s. The decision to build is binary. Modify problem (a) accordingly to reflect this. The objective is unchanged: money.
  (d) Power may be produced at this upstream facility, but it requires two things: the existence of the upstream dam; and the extra cost K_a for the construction of an aluminum refinery and associated electrical generators. These would enable power to have value V per unit produced; but at most, P_a power could be produced and used. Make these final modifications to your formulation.

28 Here is a recorded time series of the logarithm of streamflow: (5, 4, 2, 4, 5, 6, 5, 2, 3, 4). Construct a synthetic streamflow generator using a lag-1 Markov process.

29 Edwards Falls is shown in Figure 5.24, with inflow $Q = 10$. There is a farm that withdraws W units of water. The farm grows a single crop corn, X. Each unit of corn

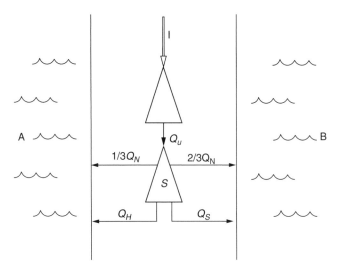

Figure 5.23. The high Lake Ganut, as in Figure 5.22, with upstream reservoir added for Problem 27c, d.

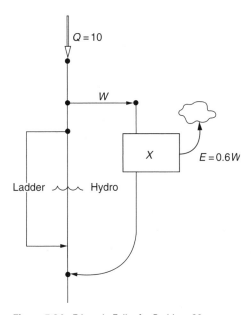

Figure 5.24. Edwards Falls, for Problem 29.

requires (5, 6, 2) units of water, land, and fertilizer. Available land is 10; available fertilizer is 7.

Downstrean there is a run-of-river hydropower facility; and a fish ladder around it that must have flow = 2 at all times.

(a) What is the upper limit of W?

(b) What is Q_h, the flow through the hydro facility?

(c) Formulate this problem as an LP, using the two decision variables W and X. Objective: Maximize $(V_x X + V_h Q_h)$.

(d) $V_h = 1$. What range of values of V_x will result in what optimal solution?

30   Two irrigation technologies are under consideration. You must choose one or the other. In one case, the production constraints would be

$$5X_1 + 2X_2 < 6 \tag{5.185}$$

$$8X_1 + 1X_2 < 6 \tag{5.186}$$

In the other case, the constraints would be

$$5X_1 + 3X_2 < 2 \tag{5.187}$$

$$9X_1 + 2X_2 < 10 \tag{5.188}$$

Implementing the second case would require a capital investment K and would result in an increase in the value of X_1 by C. There are several other constraints not shown, all unchanged by the choice of technology.

Formulate (do not solve) this problem as a Mixed-Integer-Programming problem.

6 Pollution

Contamination affects all natural resources. Our basic classification scheme uses two binary axes, and each of the four quadrants is illustrated in previous chapters without reference to contamination. But essentially, there is a third binary axis, perhaps "degradability." Simpson et al. [82], in revisiting the classic Barnett/Morse examination of finite resources [3], referred to the " 'New Scarcity' – the limitations on the environment's capacity to absorb and neutralize."

The general field of pollution is huge. Here we introduce some basic analyses, for further consideration.

The context: Waste material is loaded into a receiving medium. The material is assumed to be dangerous when aggregated. It will be accumulated, removed, transformed, and/or sequestered, but accumulation alone is not sustainable. An acceptable balance with at least one other rate is necessary if sustained loading is to occur.

The medium itself is a "resource in reverse." Separate loadings have cumulative effects on the common receiving medium. Its assimilative capacity is the rate at which aggregate loading can be received and processed with acceptable accumulation. A sustainable balance between loading and accumulation relies on the other rates mentioned: removal, transformation, sequestration. Managing these rates, and the collective loading they balance, is the resource management challenge.

6.1 BASIC PROCESSES

The basic system is illustrated in Figure 6.1. The variables are

M: the mass of material
L: the loading rate, mass/time
R: the removal rate, mass/time
V: the volume of the receiving medium
Q_i: the volumetric flow rate of fluid entering the system, volume/time
Q: the volumetric flow rate of fluid leaving the system, volume/time

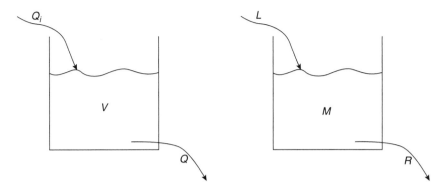

Figure 6.1. Loading, accumulation, and removal from a volume. *Left:* Hydraulic variables; *Right:* Mass variables.

In terms of these, we have the simple conservation of mass:

$$\frac{dM}{dt} = L - R \qquad (6.1)$$

$$\frac{dV}{dt} = Q_i - Q \qquad (6.2)$$

If $C \equiv M/V$ is the concentration, then we can differentiate Equation 6.1 by parts:

$$C\frac{dV}{dt} + V\frac{dC}{dt} = (L - R) \qquad (6.3)$$

$$C\left(\frac{1}{V}\frac{dV}{dt}\right) + \frac{dC}{dt} = \frac{1}{V}(L - R) \qquad (6.4)$$

Define $L' = L/V$ and similarly $R' = R/V$. Further, define the hydrodynamic convergence rate $a \equiv \left(\frac{1}{V}\frac{dV}{dt}\right)$. Then we have

$$\frac{dC}{dt} + aC = L' - R' \qquad (6.5)$$

This form exposes the dilution (L') and convergence (a) effects of the flow regime on C. Thoughout, a primed quantity indicates a mass flux normalized by V, with units of C/time.

Hereafter, we will assume that the system is in hydraulic steady state, with constant volume V and $Q = Q_i$. Hence, $a = 0$, and we define the constant system residence time τ in terms of its size relative to its throughflow rate:

$$\tau \equiv \frac{V}{Q} \qquad (6.6)$$

Figure 6.2 adds three important quantities:

B: the biomass
T: the rate of internal transformation of M, mass/time
S: the rate of sequestration of M, mass/time

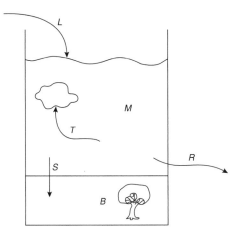

Figure 6.2. As in Figure 6.1, with biomass B at the bottom and transformation and sequestration fluxes (T, S) indicated.

There are many possible definitions of assimilative capacity $\mathcal{A}$. Here we suggest the steady-state relation between loading rate L and accumulated concentration C:

$$\mathcal{A}L = C \tag{6.7}$$

$\mathcal{A}$ in this definition will have units of inverse flow rate, time/volume. It will depend on the internal balance among rates achievable in the steady state. In more complex situations of many loadings and many measurement points, this generalizes to a matrix relation (see Section 6.4).

6.1.1 Dilution, Advection, and Residence Time

Suppose R is simply the export of C due to throughflow $Q : R = QC$. For this case, we have

$$\frac{dC}{dt} + \frac{C}{\tau} = L' \tag{6.8}$$

The loading is diluted into the constant volume V; and the removal rate is proportional to the built-up concentration, with the underlying throughflow doing the removal. C will have a steady state, where the equilibrium $R = L$ is maintained:

$$C = \tau L' \tag{6.9}$$

$$= L/Q \tag{6.10}$$

Nonequilibrium conditions will decay to this steady balance at the exponential rate $e^{(-t/\tau)}$.

The medium is unbounded in this case. Any loading can be accommodated; the built-up concentration is proportional to it. If there is a maximum $\overline{C}$ allowed, that puts an upper limit on the loading:

$$\overline{L}' = \overline{C}/\tau \tag{6.11}$$

or equivalently

$$\overline{L} = \overline{C}Q \tag{6.12}$$

This is assimilation by dilution and flushing. The entire loading rate is transferred through to the next destination, that of R.

Transformation

In addition to flushing, we need to account for transformation of M by biochemical action, T or $T' \equiv T/V$:

$$\frac{dC}{dt} = L' - R' - T' \tag{6.13}$$

and the steady-state balance is simply $L' = R' + T'$. (The reaction products are assumed benign, else they would have to be accounted for.)

The "first-order" transformation rate is by definition proportional to C:

$$T' = kC \tag{6.14}$$

Mathematically, it simply adds to the effect of advection, and the steady state is

$$L' = C \left(\frac{1}{\tau} + k \right) \tag{6.15}$$

First-order transformation supplements the advective affect; a given $\overline{C}$ will allow a larger loading $\overline{L}$.

Saturation

At large C, it is possible that T' saturates at the asymptotic ceiling rate T'_0, as the availability of C no longer limits T' and other factors (unaccounted for here) control. There are many useful forms, including the *exponential* form:

$$T'(C) = T'_0 \left[1 - \exp(-kC/T'_0) \right] \tag{6.16}$$

and the *hyperbolic* form:

$$T'(C) = kC / \left[1 + kC/T'_0 \right] \tag{6.17}$$

In both cases, there are two constants. As expressed here, they share the same features: $T'(C) = kC$ at low C and saturating at T'_0 at high C.

The quantity T'_0/k is the "half-saturation constant" C_h. Rearranging the rate expressions, we have the equivalent forms: *exponential*

$$\frac{T'(C)}{T'_0} = \left[1 - \exp(-C/C_h) \right] \tag{6.18}$$

and *hyperbolic*:

$$\frac{T'(C)}{T_0'} = \left[\frac{C/C_h}{1 + C/C_h}\right] \tag{6.19}$$

Again, both forms saturate at T_0' and have the same linear limit at low C:

$$\frac{T'(C)}{T_0'} = \frac{C}{C_h} \tag{6.20}$$

There are three parameters, k, T_0', C_h, but only two independent ones, with the third one determined from the requirement

$$C_h \equiv T_0'/k \tag{6.21}$$

The hyperbolic form approaches saturation more slowly – for example, when $C = C_h$, T' is half of its saturation value T_0'. If the exponential form is used instead, T' reaches 63% of saturation at this concentration.

Figure 6.3 shows these two forms, arranged as above to share the same limits. Many other forms are possible.

Equilibrium will occur when loading balances advection and transformation, $L' = C/\tau + T'(C)$. At low C, we will have

$$L' = C\left(\frac{1}{\tau} + k\right) \tag{6.22}$$

At high C, the limit will be

$$L' = C\left(\frac{1}{\tau}\right) + T_0' \tag{6.23}$$

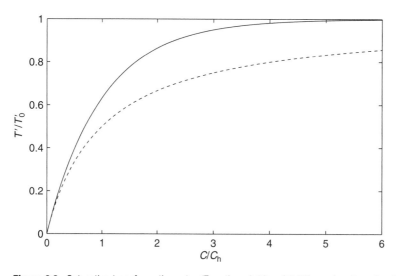

Figure 6.3. Saturating transformation rates (Equations 6.16 and 6.17) as a function of ambient concentration (exponential form, solid line; hyperbolic form, dash line). Concentration is unbounded as loading approaches the limiting value T_0'. $C_h \equiv T_0'/k$.

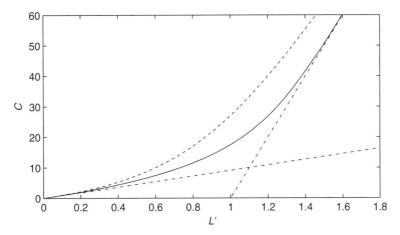

Figure 6.4. Concentration in equilibrium with steady loading rate. Advection with residence time $\tau = 100$; plus saturating transformation rate with $k = .1$, $T_0' = 1$ ($C_h = 10$). *Solid:* exponential form; *dash:* hyperbolic form. The two straight lines are the asymptotic limits from Equations 6.24 and 6.25.

Equivalently,

$$L' < L_0' : C = \tau L' / \left(1 + k\tau\right) \tag{6.24}$$

$$L' > L_0' : C = \tau \left(L' - T_0'\right) \tag{6.25}$$

with the break point L_0':

$$L_0' = \left(\frac{1 + \tau k}{\tau k}\right) T_0' \tag{6.26}$$

(see Figure 6.4).

6.1.3 Sequestration

Suppose there is an inventory of biomass B, fixed to the bottom as in Figure 6.2. Let the units of B be the same as that of M – for example, mass of a dominant element. If B consumes M, transforming it biochemically and sequestering it in its tissue, then the sequestration rate S must be the same as the biomass growth rate G:

$$\frac{dM}{dt} = -S \tag{6.27}$$

$$\frac{dB}{dt} = G \tag{6.28}$$

$$S = G \tag{6.29}$$

Let the growth rate G for the biomass be given by the logistic rate

$$G = gB\left(1 - B/B_0\right) \tag{6.30}$$

where B_0 is the biomass carrying capacity. Here we assume that at reasonable abundance, G and therefore S are not limited by M – things other than M limit, for example, sunlight, nutrients, temperature, etc. Under these conditions, B will undergo logistic growth toward B_0 at the expense of M. Logistic growth of B was elaborated in an earlier chapter.

Sequestration thus occurs only during growth, with M transferred to B. Ultimately, sequestration ceases when B's carrying capacity is reached. M transferred into the biomass is held there permanently. If we add a finite residence time τ, then continous flushing stabilizes M in steady state before and after growth/sequestration. Figure 6.5 (top panel) illustrates this process.

So far, growth is limited only by B as it approaches carrying capacity. At low abundance, M can also limit – certainly, M cannot be negative. Suppose the exponential

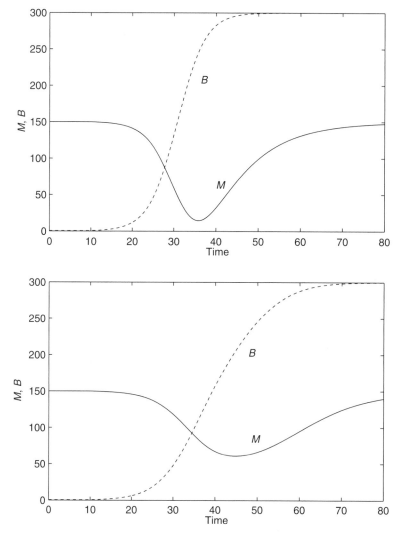

Figure 6.5. Sequestration of M by growth of B. The background steady state is stabilized by flushing at residence time $\tau = 10$; constant loading rate $L' = 15$; and biological parameters $g = .3$; $B_0 = 300$. *Top*: G insensitive to M; *bottom*: exponential M limitation on G: $g_0 = .3$; $M_h = 100$.

form governs this limitation on g:

$$g(C) = g_0 \left[1 - \exp(-C/C_h)) \right] \tag{6.31}$$

$$= g_0 \left[1 - \exp(-M/M_h)) \right] \tag{6.32}$$

Figure 6.5 (bottom panel) illustrates this dynamic. The growth is slower with less sag in M.

At low M, the limiting form of G is

$$G = \frac{g_0}{M_h} MB \left(1 - B/B_0 \right) \tag{6.33}$$

Near $B = 0$ and $M = 0$, growth has the bilinear limit:

$$G = \frac{g_0}{M_h} MB \tag{6.34}$$

In this limit, growth responds linearly to M and/or B.

Harvesting

Sequestered biomass may be harvested at rate H; in that case, it is critical to account for the harvest disposal. Figure 6.6 indicates two different destinations: export, representing permanent isolation from M; and recycling, representing return of the harvest to M. These two different fates are illustrated in the carbon case study, in which harvest recycling would represent burning and return of the combustion products to the atmosphere, while export would represent permanent isolation of the harvest from the atmosphere.

6.1.4 Bioaccumulation

Suppose a water volume V is loaded with a tracer at rate L, with residence time τ, accumulated tracer mass M, and concentration $C = M/V$. Let the bottom be

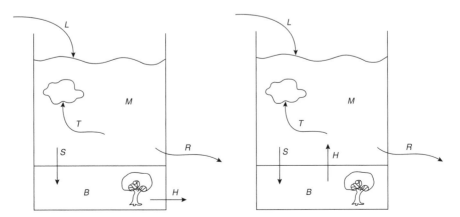

Figure 6.6. Destination of biomass harvest. *Left*: H exported; *Right*: H recycled.

occupied by a species of shellfish. There are N such animals. Each filters the ambient water, sequestering the available tracer at the rate $k_1 C$; hence, the total sequestration rate S is

$$S = k_1 CN \tag{6.35}$$

Let the total mass of tracer present in the shellfish population be M_s, which is discharged from all the shellfish together at the rate $D = D_1 + D_2$:

$$D_1 = k_2 M_s \tag{6.36}$$

$$D_2 = \mu M_s \tag{6.37}$$

D_1 represents simple recycling of M_s back into M; D_2 represents removal of M_s from the system, and/or its permanent biochemical transformation into isolated forms.

The mass balances are

$$\frac{dM}{dt} = L - S + D_1 - M/\tau \tag{6.38}$$

$$\frac{dM_s}{dt} = S - D_1 - D_2 \tag{6.39}$$

$$\frac{d(M + M_s)}{dt} = L - M/\tau - D_2 \tag{6.40}$$

In steady state, the balance of M_s (Equation 6.39) requires

$$k_1 CN = (k_2 + \mu)M_s \tag{6.41}$$

and thus

$$\left(\frac{k_1}{k_2 + \mu}\right) C = \frac{M_s}{N} \tag{6.42}$$

M_s/N is the mass of pollutant accumulated per shellfish. There is a bioconcentration factor $k_1/k_2 + \mu$. The overall steady state (Equation 6.40) requires

$$L = \frac{M}{\tau} + D_2 \tag{6.43}$$

and thus

$$L = \frac{M}{\tau} + \mu \left(\frac{k_1}{k_2 + \mu}\right) NC \tag{6.44}$$

$$= M \left(\frac{1}{\tau} + \mu \left(\frac{k_1}{k_2 + \mu}\right) \frac{N}{V}\right) \tag{6.45}$$

Rearranging, we find the relation between loading per volume L' and ambient C:

$$C = \frac{\tau L'}{1 + \tau \left(\frac{\mu k_1}{k_2 + \mu}\right) \frac{N}{V}} \tag{6.46}$$

In the limit of zero shellfish ($N = 0$), this reverts to the familiar flushing-only result (Equation 6.9).

6.2 CASE STUDY: CARBON

Suppose the M of interest is atmospheric carbon as carbon dioxide and the B is organic carbon on the surface of the earth. B undergoes logistic growth; parameters are $g_0 = .5$ and limited by $M_h = 250$; carrying capacity $B_0 = 500$. The loading is combustion effluent, at the constant rate $L = 50$, representing $1.8L$ in power production; the atmosphere is stabilized by exogenous processes leading to residence time $\tau = 10$.

   This dynamic is illustrated in Figure 6.7. Each panel begins in steady state.

1. Top: Initially, $B = 0$; it is seeded at a small increment and allowed to grow until it reaches the MSY point – for logistic growth this is $B = .5B_0$. M is sequestered into B, reaching a low value of roughly $M = 300$. At that point, harvesting begins, and thereafter B is held fixed by the harvest. The harvest is oxidized and returned to the atmosphere; power production is supplemented by the oxidized harvest: $E = 1.8(L + H)$. With $G = H$, there is no sequestration of M; it returns to its original steady state. The low value of M limits the growth and harvest, visible in the plot of g. As M recovers, there is a small effect on H and therefore E. Simulation: **AirTreeLimC**.
2. Middle: This is the same scenario, but there is an initial B that must be removed at the time of planting and oxidized. This causes M to undergo an initial step increase, and the system responds thereafter as in the top panel. Simulation: **AirTreeLimD**.
3. Bottom: Here we have initial clearing and steady harvesting of the MSY, but we remove the harvested biomass from the system. The result is no supplement to the energy production; instead, the removal of harvested biomass from the system results in lower M. Simulation: **AirTreeLimE**.

6.3 AERATION

Suppose M represents dissolved or suspended organic matter, measured in terms of the oxygen needed to convert it to inorganic reaction products. We need to account for C_B, its concentration, and C_{Ox}, the concentation of dissolved oxygen required to complete the reaction. Let the rate of the reaction be $k_1 C_B$, assuming that dissolved oxygen is readily available.

   As the reaction proceeds, oxygen is depleted from its natural equilibrium C^* leading to resupply at the reaeration rate $k_2(C^* - C_{Ox})$. The dynamic of the two dissolved reactants is

$$\frac{dC_B}{dt} = -k_1 C_B \tag{6.47}$$

$$\frac{dC_{Ox}}{dt} = -k_1 C_B + k_2(C^* - C_{Ox}) \tag{6.48}$$

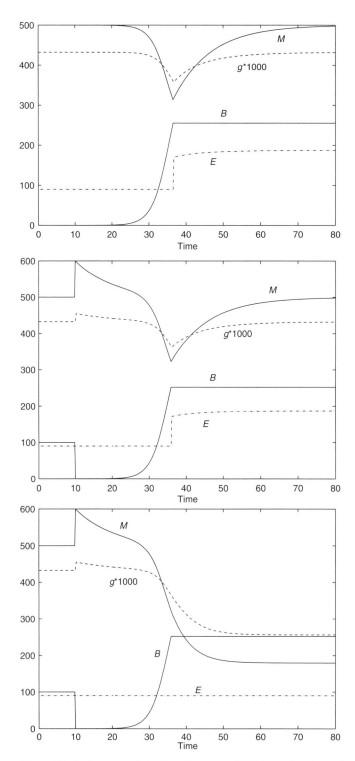

Figure 6.7. Carbon case study. *Top*: Base case, MSY harvesting, burned as fuel; *middle*: initial biomass clear/burn added; *bottom*: Harvest removed from system. *M*: atmospheric carbon; *B*: biomass; *E*: power production; *g*: logistic growth rate coefficient, limited by *M*.

If we identify the departure from equilibrium as the *oxygen deficit* $C_D \equiv (C^* - C_{Ox})$, then we have the simpler expression

$$\frac{dC_B}{dt} = -k_1 C_B \tag{6.49}$$

$$\frac{dC_D}{dt} = +k_1 C_B - k_2 C_D \tag{6.50}$$

Both C_B and C_D are ideally zero. The presence of C_B leads to the presence of C_D, which is cured ultimately by reaeration.

For constant k_1 and k_2, the solution is

$$C_B = C_{Bo} e^{-k_1 t} \tag{6.51}$$

$$C_D = C_{Do} e^{-k_2 t} + C_{Bo} \left(\frac{k_1}{k_2 - k_1} \right) \left[e^{-k_1 t} - e^{-k_2 t} \right] \tag{6.52}$$

where (C_{Bo}, C_{Do}) are the conditions at $t = 0$. Figure 6.8 illustrates this solution. It is the base case of the Streeter-Phelps dissolved-oxygen analysis. C_B decays monotonically; C_D will rise initially, peaking at t_p and then decaying monotonically. For the case $C_{Do} = 0$, the time to peak is

$$t_p = \left(\frac{1}{k_1 - k_2} \right) \ln \frac{k_1}{k_2} \tag{6.53}$$

The reaeration rate will be very sensitive to the turbulence present at the air–water interface and to the surface-to-volume ratio of the water body. The decay rate k_1 will be sensitive to the presence of oxygen, clearly shutting down at $C_{Ox} = 0$. Typically, it

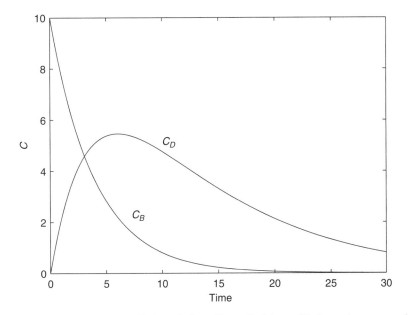

Figure 6.8. Basic Streeter-Phelps solution with constant $k_1 = .25$, $k_2 = .1$; no oxygen limitation

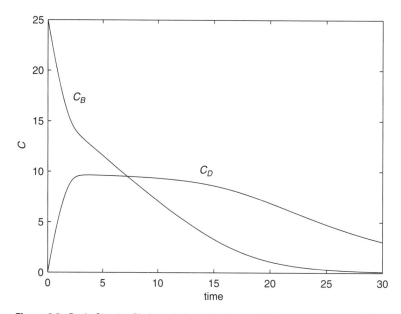

Figure 6.9. Basic Streeter-Phelps solution as in Figure 6.8, but with oxygen limitation as in Equation 6.54: $C_h = 1$ and equilibrium $C^* = 10$. The higher initial condition C_{Bo} results in higher deficit C_D, which in turn limits the reaction rate.

saturates well below C^*, but a half-saturation constant C_h will be required to limit it at very low C_{Ox}

$$k_1(C_{Ox}) = k_{1o}\left[1 - e^{-C_{Ox}/C_h}\right] \tag{6.54}$$

Figure 6.9 illustrates this effect by scaling up the initial conditions to provoke the oxygen limitation.

Program **SPbox.m** simulates this case.

Anoxia

Figure 6.9 reveals the distinctive slow recovery from anoxia, during which C_D is nearly constant at $C_D = C^*$; the reaeration rate approaches its upper limit $R^* = k_2 C^*$; and the decay and reaeration rates balance. In the low-oxygen limit, this rate balance is

$$k_{1o}\frac{C_{Ox}}{C_h}C_B = k_2 C^*$$

C_B declines at the constant rate R^*; this is clearly visible in Figure 6.9, wherein $R^* = 1$ and the anoxia persist for a period order 10–15 time units. C_{Ox} responds as its inverse:

$$C_{Ox} = \frac{k_2 C^* C_h}{k_{1o}}\frac{1}{C_B}$$

The Streeter-Phelps River

The above analysis assumes a closed system exposed to the atmosphere. Suppose a "parcel" of water is moving down a river at speed v. The simplest extension would be to follow the parcel so that the distance traveled would be $x = vt$. Hence, we would expect to see Streeter-Phelps dynamics realized in spatial patterns. Constant loading L_B (mass/time) into a uniform flow rate Q (volume/time) would initially create $C_{Bo} = L_B/Q$. Downstream, this concentration would have an increasing age $t = x/v$. The oxygen deficit would intially grow with distance, peak, and then decay; and generally, the temporal patterns displayed in Figures 6.8 and 6.9 would be mapped onto x.

For constant coefficients, we have the linear transformation

$$\left\{ \begin{array}{c} C_B \\ C_D \end{array} \right\} = \left[\begin{array}{cc} e^{-k_1 x/v} & 0 \\ \left(\frac{k_1}{k_2-k_1}\right)\left[e^{-k_1 x/v} - e^{-k_2 x/v}\right] & e^{-k_2 x/v} \end{array} \right] \left\{ \begin{array}{c} C_{Bo} \\ C_{Do} \end{array} \right\} \qquad (6.55)$$

It is convenient to summarize this as the matrix transformation of water quality $\{C\}$:

$$\{C(x)\} = [T]\{C(0)\} \qquad (6.56)$$

$[T]$ is specific to a particular segment or reach of a river. If reaches are numbered sequentially downstream, then

$$\{C\}_3 = [T]_3 \{C\}_2 \qquad (6.57)$$

$$= [T]_3 [T]_2 \{C\}_1 \qquad (6.58)$$

$$= [T]_3 [T]_2 [T]_1 \{C\}_0 \qquad (6.59)$$

and, of course,

$$\{C\}_0 = \frac{1}{Q} \{L\}_0 \qquad (6.60)$$

This concatenation of discrete transformations is relevant where distinct reaches support locally constant values of x/v, k_1, and k_2. Additional elements are needed: For example, where two tributaries Q_1 and Q_2 merge into Q_3, we have the mixing relation demanded by mass balance on an infinitesimal reach:

$$\{C\}_3 = \frac{Q_1}{Q_3} \{C\}_1 + \frac{Q_2}{Q_3} \{C\}_2 \qquad (6.61)$$

Impoundments can be handled as "reaches," with residence times as in previous analyses: $t = V/Q$ and the likelihood of small reaeration coefficients k_2.

The Streeter-Phelps Pond

At the other extreme from the river, we imagine a lake with steady hydraulic residence time τ, Streeter-Phelps dynamics internally, and loading rate L':

$$\frac{dC_B}{dt} = -k_1 C_B - \frac{1}{\tau}C_B + L' \tag{6.62}$$

$$\frac{dC_D}{dt} = -k_2 C_D - \frac{1}{\tau}C_D + k_1 C_B \tag{6.63}$$

This lake will support a steady state. Balancing the rates on the right sides, we obtain

$$C_B = \left(\frac{1}{k_1 + 1/\tau}\right) L' \tag{6.64}$$

$$C_D = \left(\frac{k_1}{k_2 + 1/\tau}\right)\left(\frac{1}{k_1 + 1/\tau}\right) L' \tag{6.65}$$

6.4 MULTIPLE LOADING

6.4.1 Lake Hitchcock

Consider the aquatic system in Figure 6.10. There are two steady pollution loadings L_1 and L_2 (mass/time) and three locations where water quality C (mass/volume) is important: a water supply intake, a state park, and a critical habitat for an endangered species. These are all linked by a complex circulation system as indicated.

Each loading has an associated *unit profile*, as shown in Table 6.1. This is the value of C at the three monitoring points that results from a unit loading rate. Each profile comprises three scalar *influence coefficients* $a_{i,j}$ – the water quality at point i that results from unit loading at point j alone.

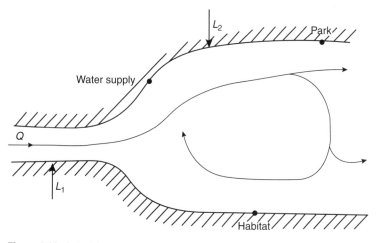

Figure 6.10. Lake Hitchcock with two loading points L_1 and L_2 and three monitoring stations.

Table 6.1. Lake Hitchcock data with two loadings and three water quality monitoring points. C^* indicates required water quality.

	Unit Profile		Status Quo	Limit	Required Improvement
	L_1	L_2	C	C^*	ΔC
Water Supply	2	0	4	≤ 3	≥ 1
Park	2	1	5	≤ 2	≥ 3
Habitat	1	3	5	≤ 2	≥ 3
Loading Status Quo	2	1			

The superposition of all three loadings results in the total water quality impact

$$L_1 \begin{Bmatrix} 2 \\ 2 \\ 1 \end{Bmatrix} + L_2 \begin{Bmatrix} 0 \\ 1 \\ 3 \end{Bmatrix} = \begin{bmatrix} 2 & 0 \\ 2 & 1 \\ 1 & 3 \end{bmatrix} \begin{Bmatrix} L_1 \\ L_2 \end{Bmatrix} = \begin{Bmatrix} C_W \\ C_P \\ C_H \end{Bmatrix} \tag{6.66}$$

The matrix form is the most compact expression of this:

$$[A]\{L\} = \{C\} \tag{6.67}$$

The columns of A are the unit profiles; the scalar entries are the influence coefficients.

The status quo loadings in Table 6.1 are $L_1 = 2$ and $L_2 = 1$; they result in the water quality status quo concentrations shown. At all monitoring locations, C exceeds the indicated requirements C^*. Reduction in L_1 and/or L_2 is necessary, so the requirement is to find L such that

$$[A]\{L\} \leq \{C^*\} \tag{6.68}$$

Let V_1, V_2 be the costs of reducing loadings (treatment). Then the cost of achieving water quality is

$$Z = V_1(2 - L_1) + V_2(1 - L_2) \tag{6.69}$$

So the minimum-cost solution is a linear programming optimization: Minimize Z subject to the constraint (Equation 6.68). Implied is that L must be nonnegative.

Clearly, minimizing Z is the same as maximizing $-Z$. And ignoring the constant, we have the equivalent objective: Maximize $V_1(L_1) + V_2(L_2)$. Effectively, maximize loadings subject to the water quality constraints.

There are likely other constraints on L. For example, it is entirely possible that an ideal loading could *increase* and be compensated for by reductions in loading elsewhere, depending on the relative costs. If this is to be ruled out, we need to constrain L beneath the status quo $\bar{L}$; in this case,

$$0 \leq L_1 \leq 2 \tag{6.70}$$

$$0 \leq L_2 \leq 1 \tag{6.71}$$

So this problem may be summarized:

$$\text{Maximize } \{V\} \cdot \{L\} \tag{6.72}$$

subject to

$$[A]\{L\} \leq \{C^*\} \tag{6.73}$$

and

$$\{0\} \leq \{L\} \leq \left\{\overline{L}\right\} \tag{6.74}$$

Effectively, maximize loadings subject to the water quality constraints and do not increase any loadings.

There is an interesting equivalent formulation. Let the *reductions* in loadings be R_1, R_2. Reworking the algebra with $L = \overline{L} - R$ and the status quo $C = A\overline{L}$, we have

$$\text{Minimize } \{V\} \cdot \{R\} \tag{6.75}$$

subject to

$$[A]\{R\} \geq \{\Delta C\} \tag{6.76}$$

and

$$\{0\} \leq \{R\} \leq \left\{\overline{L}\right\} \tag{6.77}$$

Essentially, minimize reductions subject to minimum acceptable improvements in water quality. The same matrix A of influence coefficients is necessary to describe the natural water quality/loading interactions.

Many features could be added. For example, suppose it were not possible to eliminate L_1 altogether. This would require a minimum acceptable loading $\underline{L}_1$ and, in the equivalent formulation, a maximum acceptable reduction $\overline{R}_1$.

Figure 6.11 illustrates the constraints for this problem, using the R-based formulation. The solution for R is required to be in the first quadrant ($R \geq 0$), bounded by upper limits on R (equivalently, $L = 0$). The water supply constraint is always slack; the park and habitat constraints, plus the upper limits on R, bound the feasible region. Because this problem minimizes a weighted sum of R, the optimum will lie on the inner boundary of the feasible region. There are three interesting extrema, labeled A, B, C in the figure. Each is the intersection of exactly two binding constraints:

- Point A: $R_1 = 1$, $R_2 = 1$. Water quality at the park binds, and L_2 is at its lower limit 0. This extremum will be optimal when reduction at point 1 is very costly:

$$2 < \frac{V_1}{V_2} \tag{6.78}$$

Water quality at the habitat and water supply points is better than required; those constraints are slack.

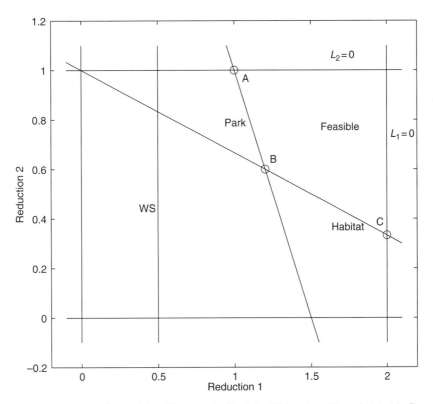

Figure 6.11. Constraints and feasible space for the Lake Hitchcock problem sketched in Figure 6.10 and Table 6.1. Three corner solutions A, B, C are highlighted, and their optimality is discussed in the text.

- Point B: $R_1 = 6/5$, $R_2 = 3/5$. Water quality at the park and the habitat points bind. This extremum will be optimal for intermediate costs:

$$\frac{1}{3} < \frac{V_1}{V_2} < 2 \tag{6.79}$$

Both polluters reduce partially; the requirements are exactly met at the park and habitat; at the water supply point, quality is better than required.

- Point C: $R_1 = 2$, $R_2 = 1/3$. Water quality at habitat point binds, and L_1 is driven to zero. This extremum will be optimal when reduction at point 2 is very costly:

$$\frac{V_1}{V_2} < \frac{1}{3} \tag{6.80}$$

Water quality at the park and water supply points exceeds requirements.

6.4.2 Wheelock-Kemeny Basin

We return to the Wheelock-Kemeny River Basin introduced in Section 5.7. Figure 6.12 shows the topography with two input flows and two reservoirs. Hydraulic aspects of this system were discussed in Section 5.7 and the related problems.

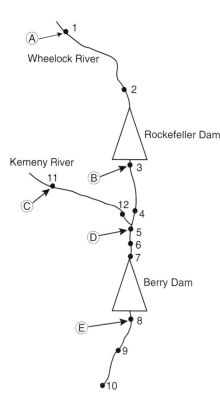

Wheelock River

Kemeny River

Rockefeller Dam

Berry Dam

Figure 6.12. Wheelock-Kemeny Basin from Figure 5.17 with pollution sources (lettered) and monitoring stations (numbered) indicated

Here we imagine a seasonal hydraulic regime comprising storage levels in both reservoirs and flows exiting them. These management variables plus the two inflows amount to six hydraulic variables. They are assumed to be in a quasi-steady state during the low-flow season, typically the limiting season for water quality when in-stream dilution is low. Loading points are illustrated in Figure 6.12 as lettered points, and numbered water quality monitoring stations are shown. We focus on steady-state analysis of oxygen deficit; Streeter-Phelps interactions among C_B and C_D are assumed to govern, as developed in Section 6.3. The problem will be the same as in the Lake Hitchcock discussion: Assuming the flow regime is fixed, adjust the pollutant loadings to meet water quality standards. All parameters are listed in Tables 6.2 and 6.3.

Table 6.4 shows a water quality influence matrix $[A]$ for this system:

$$[A]\{L\} = \{C\} \tag{6.81}$$

assuming Streeter-Phelps dynamics with $(k_1, k_2) = (0.30, 0.15)$ days^{-1} everywhere, with the sole exception that at the reservoirs $k_2 = .01$ represents reduced reaeration due to low surface turbulence and greater depth. (In dealing with these units, it is useful to note that $1 \text{ kg/m}^3 = 10^3 \text{ mg/l}$.)

The transit time along the main-stem Wheelock, excluding the reservoirs, is about nine days, comparable to the reaeration time. So we can expect C_D to be present throughout the basin. The reservoirs have very long residence times; there is little

Table 6.2. Status quo loading L and treatment cost V for the Wheelock-Kemeny water quality case. L units: 10^7 kg per two-month season. Cost units: $M/year/unit reduction in L.

Loading Point	Loading Rate	Unit Treatment Cost
A	20	0.5
B	10	0.5
C	5	0.5
D	8	0.5
E	5	0.5

Wheelock River	
Q_i	60
v	0.5
Distance from Point 1	
2	50
3	52
4	100
5	100
6	200
7	260
8	270
9	350
10	400

Table 6.3. Data for the Wheelock-Kemeny water quality case. Flow Q units: 10^8 m^3 per two-month season; distance units: km; velocity v units: m/s; storage volume S: 10^8 m^3.

Kemeny River	
Q_i	50
v	1.0
Distance from Point 11	
12	120
5	120

Rockefeller Reservoir	
S	150
Q_0	60
Berry Reservoir	
S	400
Q_0	110

Table 6.4. Influence coefficients for the Wheelock-Kemeny water quality case. Each column is the water quality profile for unit loading of BOD. Each station shows concentrations of BOD and oxygen deficit, C_B and C_D. L units: 10^7 kg per two-month season. C units: mg/l.

| Response Point | Loading Point | | | | | Status Quo |
	A	B	C	D	E	C
1:C_B	1.667	0	0	0	0	33.33
1:C_D	0	0	0	0	0	0
2:C_B	1.178	0	0	0	0	23.56
2:C_D	0.447	0	0	0	0	8.93
3:C_B	0	1.667	0	0	0	16.67
3:C_D	0.371	0	0	0	0	7.43
4:C_B	0	1.194	0	0	0	11.94
4:C_D	0.314	0.433	0	0	0	10.62
5:C_B	0	0.651	0.599	0.909	0	16.78
5:C_D	0.172	0.236	0.278	0	0	7.18
6:C_B	0	0.325	0.299	0.454	0	8.38
6:C_D	0.121	0.437	0.445	0.377	0	12.03
7:C_B	0	0.214	0.197	0.299	0	5.52
7:C_D	0.098	0.454	0.452	0.445	0	12.33
8:C_B	0	0	0	0	0.909	4.55
8:C_D	0.011	0.076	0.074	0.085	0	2.03
9:C_B	0	0	0	0	0.522	2.61
9:C_D	0.008	0.058	0.056	0.064	0.334	3.21
10:C_B	0	0	0	0	0.369	1.84
10:C_D	0.007	0.049	0.047	0.054	0.421	3.40
11:C_B	0	0	2.000	0	0	10
11:C_D	0	0	0	0	0	0
12:C_B	0	0	1.319	0	0	6.59
12:C_D	0	0	0.611	0	0	3.05
Loadings	20	10	5	8	5	Status Quo

reaeration there. But organic decay will run to completion in them, converting all entering C_B to C_D. These effects can be seen in the unit profiles recorded in Table 6.4.

Table 6.4 also records the status quo water quality profile of the river system, obtained by a weighted sum of the unit profiles. A typical value of oxygen saturation might be $10\,mg/l$; hence, it is clear that 4, 6, and 7 are close to anoxic (deficit near 10 mg/l) and leading candidates for improvement. Other points of concern include point 10: It is the point of export, and if another political or water management unit is downstream, we imagine a requirement governing C_D and C_B there.

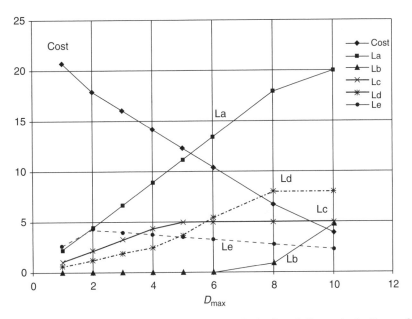

Figure 6.13. Optimal results for Wheelock-Kemeny Basin. D_{max} is the required ceiling on C_D at all stations; in addition, at station 10, C_B and C_D are required to be below 2 mg/l. All unit treatment costs are equal, $V_i = 0.5$.

Treatment costs V at all loading points are assumed to be equal. C_D is required to be below the same value D_{max} at all stations. In addition, at station 10, both C_B and C_D are required to be at or below 2 mg/l. Optimal (least-cost) loadings appear in Figure 6.13.

The optimal water quality profile for one case, $D_{max} = 5$, is shown in Table 6.5. There is improvement at all stations over the untreated status quo. The C_D constraint is binding at stations 2, 7, and 10; station 6 is very close to the limit but still slightly slack.

Program **StreeterPhelpsInfluence.xls** computes the influence coefficients for this case. Program **StreeterPhelpsLP62d.xls** performs the optimization.

6.5 RECAP

Early analyses of regional pollution recognized the common-pool nature of the problem. Dorfman et al. [21, 20] and Kneese and Bower [50] summarized early formulations that blended economic and environmental analysis, and these formulations remain useful.

Since those publications, there have been significant advances in describing receiving media in terms of basic transport processes. There are many works – for example, Lynch [56] – that place pollution in a regional and geophysical context, using a basic partial differential equation framework. Concurrently, many texts have emerged emphasizing basic pollution abatement processes; examples

Table 6.5. Wheelock-Kemeny water quality profile using original (untreated) and least-cost loadings for $D_{max} = 5$ and C_D and C_B both ≤ 2 mg/l at station 10

	Untreated	Optimal Treatment
1:C_B	33.33	18.66
1:C_D	0	0
2:C_B	23.56	13.19
2:C_D	8.93	5.00
3:C_B	16.67	0
3:C_D	7.43	4.16
4:C_B	11.94	0
4:C_D	10.62	3.52
5:C_B	16.78	6.34
5:C_D	7.18	3.31
6:C_B	8.38	3.17
6:C_D	12.03	4.97
7:C_B	5.52	2.09
7:C_D	12.33	5.00
8:C_B	4.55	3.21
8:C_D	2.03	0.81
9:C_B	2.61	1.84
9:C_D	3.21	1.79
10:C_B	1.84	1.30
10:C_D	3.40	2.00
11:C_B	10	10.00
11:C_D	0	0
12:C_B	6.59	6.59
12:C_D	3.05	3.05

include Tchobanoglous and Schroeder [87], Tchobanoglous et al. [86], Hammer and Hammer [37], HDR Engineering [45], and many others, often reflecting active professional involvement. Significant is the attention being paid to sediments (e.g., Boudreau [5]) and benthic communities generally (e.g., Di Toro [19]). Current water quality applications have reached a high level of modeling sophistication (Thomann and Mueller; Chopra); and have grown in scope to include the marine environment; examples include Fitzpatrick [26] and Nassauer et al. [67]. Most obvious is the increasing scale of pollution analysis, consistent with that of human industry. It is no surprise that the concurrent advances in networking, remote sensing, and computational power have been coupled with widespread public concern for this activity.

There has been growing concern about atmospheric carbon dioxide for over a half-century. Contemporary resources include a good description of the geophysical context (Houghton [43]); a continuing international scientific review by the Intergovernmental Panel on Climate Change (IPCC [46]); a high-visibility economic analysis (Stern, [84]); and a fresh look at institutional coordination issues (Sandler [79]). It is likely that national and international activities addressing this problem will be coordinated in the immediate future and that planetary-scale experiments will be monitored with the stunning advances in computation, planetary observation, and networking already in coordinated use. The ocean, a key player in long-term carbon dynamics, remains a frontier for examination in this and myriad other matters.

Rachel Carson, in the 1961 preface to her 1950 classic *The Sea Around Us* (Oxford University Press) [7], concludes:

The truth is that disposal has proceeded far more rapidly than our knowledge justifies.

6.6 PROGRAMS

StreeterPhelpsInfluence.xls: develops the influence coefficients for the Wheelock-Kemeny Basin.

StreeterPhelpsLP62D.xls: performs the optimization presented in the text for the Wheelock-Kemeny Basin.

SPbox.m: implements a generic Streeter-Phelps analysis of a closed system.

AirTree.m: is a generic illustration of the Carbon cycling between atmosphere and biomass: The three cases in Figure 6.7 are simulated in the series **AirTreeLimC.m, AirTreeLimD.m, AirTreeLimE.m**

Assim.m: explores the different limiting forms of assimilation rate.

6.7 PROBLEMS

1 Consider a hydraulically closed system: $Q = 0$, $\tau = \infty$, in steady state; $T'(C) = L'$. Develop the steady-state relation between loading rate L' and concentration C for two cases of saturating transformation: (a) hyperbolic (Equation 6.17); and (b) exponential (Equation 6.16). Compare both to that for the linear, first-order transformation rate (Equation 6.15).

2 A forest at carrying capacity is to be put into biofuel production. This will require instant implementation of constant harvesting at the maximum sustainable yield (MSY) and oxidation of the entire harvest. All parameters are as in the carbon case study (Section 6.2).

(a) What are the steady-state values of B, M, H, and E before harvesting begins?

(b) What will the steady-state values be after this change is implemented?

(c) Simulate the transition and compare with the relevant parts of the carbon case study.

3   Continuation of Problem 2. Harvesting is in equilibrium at MSY. A new government policy is implemented that mandates that harvest be changed instantaneously such that B be cut back to $.25B_0$ and all of the harvest be burned as fuel. L is constant throughout.

    (a) What changes have to happen instantly?

    (b) What is the new steady state?

    (c) Simulate and plot the transition.

After 20 years, the policy is again changed to eliminate the fuel production and instead sequester the harvest by sea disposal.

    (d) What changes have to happen instantly?

    (e) What is the new steady state?

    (f) Simulate and plot the transition.

4   Sequestration into a logistic biomass, in the limit of low M, is described in Equation 6.33. Express the steady-state balance among loading, sequestration, and flushing. Use this to derive a steady-state expression for M in terms of L and B.

5   The same as Problem 4, but assume that both M and B are low (Equation 6.34).

6   A lake has volume $V = 5,000$ and steady throughflow $Q = 1,000$. It is loaded by pollutant mass at the rate $L = 100$ and contains M mass of pollutant.

The lake bottom is home to an aquatic species of plant, biomass B, that has logistic growth $G = gB(1 - B/B_0)$ with $g = 30$ and $B_0 = 200$. Plant uptake of the pollutant is at the rate σG, with $\sigma = 0.1$. Plants are harvested at the constant rate H. Half of the harvest is exported; the rest is dispersed back into the lake.

    (a) What is the residence time in the lake? At what rate will the pollutant be removed by flushing?

    (b) It is observed that B is steady at 60% of carrying capacity. What is G? What is H?

    (a) What is the steady-state value of M?

7   Equation 6.53 gives the Streeter-Phelps time to peak oxygen deficit t_p, assuming the deficit is zero initially. Relax this assumption and derive the time to peak for the more general case. Show that the result recovers the simpler result when $C_{Do} = 0$.

8   For a Streeter-Phelps pond, for constant parameters τ, k_1, k_2, L', and oxygen saturation C^*: What loading will lead to total in the steady state?

9   For a Streeter-Phelps pond, for constant parameters and negligible throughflow (τ infinite): At low oxygen C_{Ox}, decay will be limited. Assume that the limiting form is $k_1 = k_{1o}C_{Ox}/C_h$, where C_h is the half-saturation constant (the low-oxygen limit of Equation 6.54). Find expressions for C_B, C_D, and C_{Ox} in terms of L'. Compare with the fully-saturated ($k_1 = k_{1o}$) case and indicate the range of validity for extremes.

10   A lake is loaded with BOD, as shown in Figure 6.14. The loading is fully saturated with oxygen. The lake volume is huge and essentially isolated from the atmosphere with

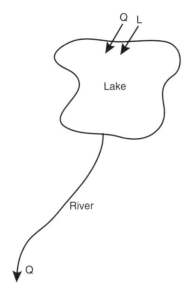

Figure 6.14. Upland lake and stream for Problem 10.

zero reaeration. A river exits the lake. Streeter-Phelps dynamics pertain, with $k_1 = .2$ and $k_2 = .05$ in the river. Velocity in the river is 0.3. Everything is in steady state.

What is the water quality downstream of the lake, as a function of distance?

11 For the Lake Hitchcock example: Evaluate the slack in each of the three water quality constraints for all possibilities of V_1/V_2.

12 In the Lake Hitchcock example: the water supply constraint never binds. Suppose C^* at that point were reduced to (a) 2; (b) 1.5; and (c) 1. For each case, sketch the feasible space and find the range of V_1/V_2 for which this constraint binds under optimal loading.

13 Implement the Lake Hitchcock example in a linear programming solver. Confirm the analysis given in the text.

14 In the Lake Hitchcock example:
  (a) Suppose $(V_1, V_2) = (25, 10)$. What is the value (money) to be gained by an increase in C^* by 0.1 at each of the three water quality monitoring points separately? What (if any) changes in loadings would occur in each case? Explain.
  (b) The same but $(V_1, V_2) = (15, 12)$.
  (c) The same but $(V_1, V_2) = (5, 20)$.
  (d) The same but $(V_1, V_2) = (45, 20)$.
  (e) The same but $(V_1, V_2) = (10, 20)$.

15 Consider the Wheelock-Kemeny Basin as presented. It is a concern that the residence time in Lake Rockefeller is too large and that anoxic waters are exiting it.
  (a) What (lower) value of S will result in a transit speed equivalent to the rest of the Wheelock? (No changes are required in any of the other hydraulics.)

(b) Using this reduced value of S, and assuming normal reaeration in the shallow reservoir ($k_2 = .15$ days^{-1}): What will the status quo be at point 3? Compare with the values prior to the change, and explain.

16 Consider the Wheelock-Kemeny Basin as presented. The Upper Valley extends from the Wheelock headwaters downstream to the confluence with the Kemeny. Formulate the Upper Valley problem with two sources and four observation points, 1 through 4.

    (a) Identify the unit responses and the influence matrix.

    (b) Formulate in terms of the loading rates L.

    (c) Repeat the formulation in terms of R.

17 Continuation of Problem 16 for the Upper Valley:

    (a) Implement the Upper Valley optimizaton, and prepare the analogue of Figure 6.13.

    (b) Prepare a companion plot of the shadow prices of all water quality constraints as a function of D_{max}.

18 Continuation of Problem 16. Change the storage in Lake Rockefeller to that found in Problem 15.

    (a) Implement the optimization of the Upper Valley with this change and prepare the analogue of Figure 6.13.

    (b) Comment on the results, and explain similarities/differences with the results of Problem 16.

19 For the Wheelock-Kemeney: consider the base case as standard, as displayed in the text: $C_D \leq 5$ everywhere, plus the requirement at station 10 that C_B and C_D both be ≤ 2. It is proposed that the Lower Valley (stations 5 through 10) be changed to $C_D \leq 6$ and the special requirements at station 10 be removed. Resolve and compare to the base case presented in the text.

20 For the Wheelock-Kemeny: The Lower Valley, stations 5 through 10, has seceeded. It is desired that $C_D \leq 4$ everywhere. The "export" requirement that C_B and C_D both be ≤ 2 at station 10, is retained.

    (a) Optimize the Lower Valley. Consider all loadings A through E to be variable, but those in the Upper Valley (A and B) and on the Kemeny (C) to cost nothing. What program is optimal?

    (b) You are negotiating a treaty with the upstream riparians. The treaty will specify water quality at points 4 and 12. What do you recommend on behalf of the Lower valley and why?

21 The Kemeny Valley is relatively undeveloped and will likely develop an important settlement halfway between stations 11 and 12, with both monitoring (Station 13) and loading (L_F) there.

    (a) Develop the Streeter-Phelps influence coefficients for this point.

(b) There is no status quo loading at point F. State clearly how that should be handled in the objective function, from the point of view of the Kemeny Valley Development Authority (KVDA).

(c) Optimize the Kemeny Valley alone, assuming a treaty obligation that both C_B and C_D not exceed 4 mg/l at Station 12, plus a domestic requirement that C_D be kept below D_{max} within the Kemeny. Plot results as a function of D_{max}.

(d) Summarize a report to the KVDA recommending domestic water quality requirements.

Generating Random Numbers

A.1 UNIFORM DEVIATE

The most common random-number generator produces the *uniform deviate*. This distribution is uniform in the interval [0, 1]. In Matlab, the function is *rand()* with no arguments. *RAND()* is the equivalent in Excel. These are pseudo-random generators. Language-specific procedures are typically needed to get these generators started (the seeding problem) in order to help avoid the associated risks. In some cases, the seeding is handled automatically, invisible to the user, by a machine clock call or other unpredictable metric. A good and practical discussion of algorithms for generating the uniform deviate can be found in Press et al. [72]. The procedures described there are recommended if routine scientific use is to be made of random-number generators.

The mean of the uniform deviate $\mathcal{U}$ is by inspection 0.5; its variance is $\frac{1}{12}$.

$$\overline{\mathcal{U}} = \frac{1}{2} \tag{1}$$

$$\sigma_{\mathcal{U}}^2 = \overline{(\mathcal{U} - \overline{\mathcal{U}})^2} = \frac{1}{12} \tag{2}$$

We can manufacture other uniform distributions by the simple transformation

$$\mathcal{V} = a(\mathcal{U} + b) \tag{3}$$

The factor b shifts the mean, while a is a scaling factor. The mean and variance are

$$\overline{\mathcal{V}} = a(\overline{\mathcal{U}} + b) \tag{4}$$

$$\sigma_{\mathcal{V}}^2 = a^2 \sigma_{\mathcal{U}}^2 \tag{5}$$

GAUSSIAN DEVIATE

If a normal, or Gaussian, distribution $\mathcal{N}$ is wanted, it can be generated from the uniform distribution $\mathcal{U}$ with the *Box-Muller transformation*:

$$\mathcal{N}_1 = \sqrt{-2\ln(\mathcal{U}_1)}\cos(2\pi\mathcal{U}_2) \tag{6}$$

$$\mathcal{N}_2 = \sqrt{-2\ln(\mathcal{U}_1)}\sin(2\pi\mathcal{U}_2) \tag{7}$$

Two values $(\mathcal{N}_1, \mathcal{N}_1)$ are generated from the two values $(\mathcal{U}_1, \mathcal{U}_2)$. Both will be members of the Gaussian distribution with zero mean, unit variance: $\mathcal{N}(0, 1)$.

$$\overline{\mathcal{N}} = 0 \tag{8}$$

$$\overline{(\mathcal{N} - \overline{\mathcal{N}})^2} = 1 \tag{9}$$

A Gaussian distribution with mean μ, variance σ^2 is gotten by shifting and scaling $\mathcal{N}$:

$$N = \sigma\mathcal{N} + \mu \tag{10}$$

The Gaussian deviate is popular enough to be supported in high-level languages – for example, in Matlab it is $randn(\mu, \sigma)$.

AUTOCORRELATED SERIES

The distributions above have zero autocorrelation. An autocorrelated sequence X_k can be constructed with the generator

$$X_k = \overline{X} + \rho\left(X_{k-1} - \overline{X}\right) + \sigma\sqrt{1 - \rho^2}\mathcal{N}_k \tag{11}$$

with $\mathcal{N}_k$ always drawn from the same uncorrelated distribution $\mathcal{N}(0, 1)$ with zero mean, unit variance. Accordingly, there are three parameters in this generator:

$\overline{X}$, the mean value of the series X_k
σ^2, the unconditional variance of the series X_k
ρ, the autocorrelation coefficient of the series X_k:

$$\rho = E(X_{k+1}X_k)/\sigma^2 \tag{12}$$

Synthetic streamflow sequences commonly employ the lognormal transformation (Loucks et al. [53]):

$$X_k = \ln(Q_k - \tau) \tag{13}$$

$$Q_k = \tau + \exp(X_k) \tag{14}$$

Here X is the autocorrelated series generated as above with the three parameters $(\overline{X}, \sigma_x, \rho_x)$. A fourth parameter τ represents the absolute minimum flow. The moments of Q are

$$\overline{Q} = \tau + \exp\left(\overline{X} + \frac{1}{2}\sigma_x^2\right) \tag{15}$$

$$\sigma_Q^2 = \left[\phi^2\right]\left[\exp\left(2\overline{X} + \sigma_x^2\right)\right] \tag{16}$$

with $\phi^2 \equiv \left[\exp\left(\sigma_x^2\right) - 1\right]$. This distribution will be skewed – that is, have a nonzero third moment:

$$\gamma_Q = \frac{E(Q - \overline{Q})^3}{\sigma_Q^3} = 3\phi + \phi^3 \tag{17}$$

For a more complete discussion of synthetic streamflow, see Loucks et al. [53].

A.4 WAITING TIME

Distributions other than the uniform and Gaussian distributions are important. It is common to generate these distributions by mapping the uniform deviate. Among the most common is the exponential, or waiting time, distribution, which has the probability density function

$$p(y) = \lambda e^{-\lambda y} \tag{18}$$

y is always positive, with mean $\overline{y} = \frac{1}{\lambda}$ and variance $\frac{1}{\lambda^2}$. This distribution is skewed toward low values, with median $= .693\overline{y}$; the probability that $y < \overline{y}$ is .632.

The generator for this distribution is

$$y = \frac{-\ln(\mathcal{U})}{\lambda} \tag{19}$$

with $\mathcal{U}$ the uniform deviate (Press et al. [72]).

A.5 SOURCES

For a useful general introduction to random numbers, see Press et al. [72]. For a useful text on ecological statistics, see Gotelli and Ellison [33]. Mangel offers a "biologist's toolbox" [60]. Hilborn and Mangel [38] provide a general discussion of modeling random processes in ecology. Pioneering work by Fiering [24, 25] was done in water resources; and a good contemporary reference in that field is Loucks et al. [53]. Hillier and Lieberman [40] provide an excellent introduction to the use of random processes in queueing theory. Editors Okubo and Levin [69] provide a through examination of diffusion processes in living systems. Cressie [15] gives a thorough review of statistical processes, and excellent depth in linear algebra is available in Demmel [18] and Trefethen and Bau [88].

BIBLIOGRAPHY

[1]   E. S. Allman and J. A. Rhodes. *Mathematical Models in Biology: An Introduction.* Cambridge University Press, 2004. 370 pp.

[2]   ASCE/UNESCO. *Sustainability Criteria for Water Resource Systems.* (Report of the ASCE Task Committee and UNESCO IHP IV Project M-4.3; D. P Loucks, chair.) American Society of Civil Engineers, 1998.

[3]   H. Barnett and C. Morse, editors. *Scarcity and Growth: The Economics of Natural Resource Availability.* Johns Hopkins University Press, 1963. 288 pp.

[4]   J. J. Bogardi and Z. W. Kundzewicz. *Risk, Reliability, Uncertainty, and Robustness of Water Resources Systems.* Cambridge University Press, 2005. 220 pp.

[5]   B. Boudreau. *Diagenic Models and Their Implementation.* Springer, 1997. 414 pp.

[6]   N. F. Britton. *Essential Mathematical Biology.* Springer-Verlag, 2003. 335 pp.

[7]   R. Carson. *The Sea Around Us.* Oxford University Press, 1961. 1961 Introduction to the 1950 edition.

[8]   H. Caswell. *Matrix Population Models: Construction, Analysis, Interpretation.* Sinauer Associates, Sunderland, MA, 2nd ed. 2001. 722 pp.

[9]   C. W. Clark. *Mathematical Bioeconomics: The Optimal Management of Renewable Resources,* 2nd ed. John Wiley, 1990. 386 pp.

[10] C. W. Clark. *The Worldwide Crisis in Fisheries: Economic Models and Human Behaviour.* Cambridge University Press, 2006. 263 pp.

[11] C. W. Clark and M. Mangel. *Dynamic State Variable Models in Ecology: Methods and Applications.* Oxford University Press, 2000. 289 pp.

[12] Jon M. Conrad. *Resource Economics.* Cambridge University Press, 1999. 213 pp.

[13] Jon M. Conrad and Colin W. Clark. *Natural Resource Economics: Notes and Problems.* Cambridge University Press, 1987. 231 pp.

[14] R. Cornes and T. Sandler. *The Theory of Externalities, Public Goods and Club Goods,* 2nd ed. Cambridge University Press, 1996. 590 pp.

[15] N. A. C. Cressie. *Statistics for Spatial Data.* John Wiley, 1991. 900 pp.

[16] J. M. Cushing. *An Introduction to Structured Population Dynamics.* Society of Industrial and Applied Mathematics, 1998. 193 pp.

[17] D. DeAngelis and L. Gross, editors. *Individual-Based Models and Approaches in Ecology* (Coastal and Estuarine Studies). Chapman and Hall, 1992. 525 pp.

[18] J. W. Demmel. *Applied Numerical Linear Algebra.* Society of Industrial and Applied Mathematics, 1997. 419 pp.

[19] D. DiToro. *Sediment Flux Modeling.* Wiley, 2001. 448 pp.

[20]   R. Dorfman and N. S. Dorfman, editors. *Economics of the Environment: Selected Readings*. Norton, 1977. 494 pp.

[21]   R. Dorfman, H. D. Jacoby, and H. A. Thomas, editors. *Models for Managing Regional Water Quality*. Harvard University Press, 1972. 453 pp.

[22]   W. Ver Eecke. *An Anthology Regarding Merit Goods: The Unfinished Ethical Revolution in Economic Theory*. Purdue University Press, 2007. 732 pp.

[23]   W. Fennel and T. Neumann. *Introduction to the Modeling of Marine Ecosystems*. Elsevier, 2004. 297 pp.

[24]   M. B. Fiering and B. B. Jackson. *Synthetic Streamflows*. American Geophysical Union, 1971. 98 pp.

[25]   M. B. Fiering. *Streamflow Synthesis*. Harvard University Press, 1967. 139 pp.

[26]   J. Fitzpatrick. "Assessing Skill of Estuarine and Coastal Eutrophication Models for Water Quality Managers." *Journal of Marine Systems*, MARSYS-01645, 17 pp (online 2008) Elsevier.

[27]   W. M. Getz and R. G. Haight. *Population Harvesting: Demographic Models of Fish, Forest and Animal Resources*. Princeton University Press, 1989. 391 pp.

[28]   P. H. Gleick. "The Human Right to Water." In P. H. Gleick, editor, *The World's Water 2000–2001: Biennial Report on Freshwater Resources*, pp. 1–17. Washington, DC: Island Press, 2000. 300 pp.

[29]   P. H. Gleick. "Water for Food: How Much Will Be Needed?" In P. H. Gleick, editor, *The World's Water 2000–2001: Biennial Report on Freshwater Resources*, pp. 63–92. Washington, DC: Island Press, 2000. 300 pp.

[30]   P. H. Gleick. "The Millennium Development Goals for Water: Crucial Objectives, Inadequate Commitments." In P. H. Gleick, editor, *The World's Water 2004–2005: Biennial Report on Freshwater Resources*, pp. 1–15. Washington, DC: Island Press, 2004. 262 pp.

[31]   P. H. Gleick, editor. *The World's Water 2006–2007*. Island Press, 2006. 368 pp. This is a biennial series orignated in 1998–99.

[32]   H. S. Gordon. "The Economic Theory of a Common-Property Resource: The Fishery." *Journal of Political Economy*, Vol. 62, No. 2 (Jan. 1954): 124–142.

[33]   N. J. Gotelli and A. M. Ellison. *A Primer of Ecological Statistics*. Sinauer Associates, Sunderland, MA, 2004. 510 pp.

[34]   R. Q. Grafton, J. Kirkley, T. Kompas, and D. Squires. *Economics for Fisheries Management* (Ashgate Studies in Environmental and Natural Resource Economics). Ashgate, 2006. 161 pp.

[35]   R. C. Griffin. *Water Resource Economics: The Analysis of Scarcity, Policies, and Projects*. MIT Press, 2006. 402 pp.

[36]   V. Grimm and S. Railsback. *Individual-Based Modeling and Ecology* (Princeton Studies in Theoretical and Computational Biology). Princeton University Press, 2005. 428 pp.

[37]   M. Hammer and M. Hammer, Jr. *Water and Wastewater Technology*, 6th ed. Prentice-Hall, 2008. 553 pp.

[38]   R. Hilborn and M. Mangel. *The Ecological Detective: Confronting Models with Data*. Princeton University Press, 1997. 315 pp.

[39]   R. Hilborn and C. J. Walters. *Quantitative Fisheries Stock Assessment*. Routledge Chapman Hall (2004 printing by Kluwer Academic Publishers), 1992.

[40]   F. Hillier and G. Lieberman. *Introduction to Operations Research*, 8th ed. McGraw-Hill, 2005. 1061 pp.

[41]   F. C. Hoppensteadt. *Mathematical Methods of Population Biology*. Cambridge University Press, 1982. 149 pp.

[42]   H. Hotelling. "The Economics of Exhaustible Resources." *Journal of Political Economy*, Vol. 39, No. 2 (April 1931): 137–175.

[43]   J. Houghton. *Global Warming: The Complete Briefing*. Cambridge University Press, 2004. 351 pp.

[44]   M. N. Hufschmidt and M. B. Fiering. *Simulation Techniques for Design of Water-Resource Systems*. Harvard University Press, 1966. 212 pp.

[45]   HDR Engineering. *Handbook of Public Water Systems*, 2nd ed. Wiley, 2001. 1136 pp.

[46]   Intergovernmental Panel on Climate Change. *Climate Change 2007*. Cambridge University Press, 2007. In three volumes: (I) *The Physical Science Basis;* (II) *Impacts, Adaptation and Vulnerability;* (III) *Mitigation of Climate Change.*

[47]   I. Kaul, P. Conceição, K. Le Goulven, and R. Mendoza, editors. *Providing Global Public Goods: Managing Globalization*. United Nations Development Programme/Oxford University Press, 2003. 646 pp.

[48]   I. Kaul and P. Conceição, editors. *The New Public Finance*. United Nations Development Programme/Oxford University Press, 2006. 664 pp.

[49]   I. Kaul, I. Grunberg, and M. Stern, editors. *Global Public Goods: International Cooperation in the 21st Century*. United Nations Development Programme/Oxford University Press, 1999. 546 pp.

[50]   A. V. Kneese and B. T. Bower. *Environmental Quality and Residuals Management*. Johns Hopkins University Press, 1979. 337 pp.

[51]   D. P. Loucks and J. S. Gladwell. *Sustainability Criteria for Water Resource Systems*. Cambridge University Press, 1999. 154 pp.

[52]   D. P. Loucks, J. R. Stedinger, and D.A. Haith. *Water Resource Systems Planning and Management*. Prentice-Hall, 1981. 559 pp.

[53]   D. P. Loucks, E. VanBeek, J. R. Stedinger, J. P. M. Dijkman, and M. T. Villars. *Water Resource Systems Planning and Management*. UNESCO, 2005. 680 pp.

[54]   R. G. Lough, E. A. Broughton, L. J. Buckley, L. S. Incze, K. Pehrson Edwards, R. Converse, A. Aretxabaleta, and F. E. Werner. "Modeling Growth of Atlantic Cod Larvae on the Southern Flank of Georges Bank in the Tidal-Front Circulation during May 1999." *Deep-Sea Research*, Pt. 2: *Topical Studies in Oceanography*, Vol. 53, Nos. 23–24 (Nov. 2006): 2771–2788.

[55]   R. G. Lough, L. J. Buckley, F. E. Werner, J. A. Quinlan, and K. P. Edwards. A General Biophysical Model of Larval Cod (Gadus Morhua) Growth Applied to Populations on Georges Bank. *Fisheries Oceanography*, Vol. 14, No. 4 (July 2005): 241–262.

[56]   D. R. Lynch. *Numerical Partial Differential Equations for Environmental Scientists and Engineers*. Springer, 2004. 388 pp.

[57]   D. C. Major. *Multi-Objective Water Resource Planning*. American Geophysical Union, 1977. 81 pp.

[58]   M. Mangel and C. W. Clark. *Dynamic Modeling in Behavioral Ecology*. Princeton University Press, 1988. 308 pp.

[59]   M. Mangel. *Decision and Control in Uncertain Resource Systems*. Academic Press, 1985. 255 pp.

[60]   M. Mangel. *The Theoretical Biologist's Toolbox*. Cambridge University Press, 2006. 375 pp.

[61]   S. A. Marglin. *Public Investment Criteria*. MIT Press, 1967. 81 pp.

[62]   D. J. McGillicuddy, D. M. Anderson, D. R. Lynch, and D. W. Townsend. "Mechanisms Regulating the Large-Scale Seasonal Fluctuations in *Alexandrium fundyense* in the Gulf of Maine: Results from a Physical-Biological Model." *Deep-Sea Research II*, Pt. 2, Vol. 52, pp. 2698–2714.

[63]   S. Meier. *The Economics of Non-Selfish Behaviour*. Edward Elgar, 2006. 168 pp.

[64]   C. Miller, D. Lynch, F. Carlotti, W. Gentleman, and C. Lewis. "Coupling of an Individual-Based Population Dynamics Model for Stocks of *Calanus Finmarchicus* with a Circulation Model for the Georges Bank Region. *Fisheries Oceanography*, Vol. 7, Nos. 3/4, (1998): 219–234.

[65]   J. D. Murray. *Mathematical Biology I: An Introduction*, 3rd ed. Cambridge University Press, 2002. 551 pp.

[66]   R. A. Musgrave and P. B. Musgrave. *Public Finance in Theory and Practice*, 5th ed. McGraw-Hill, 1989. 627 pp.

[67]   J. I. Nassauer, M. V. Santelmann, and D. Scavia, editors. *From the Corn Belt to the Gulf: Societal and Environmental Implications of Alternative Agricultural Futures.* Resources for the Future, 2007. 219 pp.

[68]   P. A. Neher. *Natural Resource Economics Conservation and Exploitation.* Cambridge University Press, 1990. 360 pp.

[69]   A. Okubo and S. Levin, editors. *Diffusion and Ecological Problems: Modern Perspectives* (Interdisciplinary Applied Mathematics, 14). Springer, 2001. 467 pp.

[70]   E. Ostrom, R. Gardner, and J. Walker. *Rules, Games and Common-Pool Resources.* University of Michigan Press, 1994. 369 pp.

[71]   E. Ostrom. *Governing the Commons: The Evolution of Institutions for Collective Action.* Cambridge University Press, 1990. 280 pp.

[72]   W. H. Press, B. P. Flannery, S. A. Teukolsky, and W. T. Vetterling. *Numerical Recipes: The Art of Scientific Computing.* Cambridge University Press, 1986. 818 pp.

[73]   J. Proehl, D. Lynch, D. McGillicuddy, and J. Ledwell. "Modeling Turbulent Dispersion on the North Flank of Georges Bank Using Lagrangian Particle Methods. *Continental Shelf Research*, Vol. 25, Nos. 7/8 (May 2005): 875–900.

[74]   T. J. Quinn and R. B. Deriso. *Quantitative Fish Dynamics.* Oxford University Press, 1999. 542 pp.

[75]   E. Renshaw. *Modelling Biological Populations in Space and Time.* Cambridge University Press, 1991. 403 pp.

[76]   S. A. Salman and D. B. Bradlow. *Regulatory Frameworks for Water Resources Management: A Comparative Study.* IBRD, 2006. 198 pp.

[77]   S. A. Salman and S. McInerney-Lankford. *The Human Right to Water: Legal and Policy Dimensions.* IBRD, 2004. 180 pp.

[78]   P. Sammarco and M. Heron, editors. *The Bio-Physics of Marine Larval Dispersal.* Number 45 in Coastal and Estuarine Studies. American Geophysical Union, 1994. 303 pp.

[79]   T. Sandler. *Global Collective Action.* Cambridge University Press, 2004. 299 pp.

[80]   A. D. Scott. "The Fishery: The Objective of Sole Ownership." *Journal of Political Economy*, Vol. 63, No. 2 (April 1955): 116–124.

[81]   I. A. Shiklomanov and J. C. Rodda. *World Water Resources at the Beginning of the 21st Century.* Cambridge University Press, 2004. 435 pp.

[82]   R. D. Simpson, M. A. Toman, and R. U. Ayres. Introduction: "The 'New Scarcity' ". In R. D. Simpson, M. A. Toman, and R. U. Ayres, editors, *Scarcity and Growth Revisited*, pp. 1–32. Resources for the Future, 2005. 320 pp.

[83]   R. D. Simpson, M. A. Toman, and R. U. Ayres, editors. *Scarcity and Growth Revisited.* Resources for the Future, 2005. 320 pp.

[84]   N. Stern. *The Economics of Climate Change: The Stern Review.* Cambridge University Press, 2007. 692 pp.

[85]   J. Stewart. *Calculus*, 4th ed. Brooks-Cole, 1999.

[86]   G. Tchobanoglous, F. Burton, and D. Stensel. *Wastewater Engineering*, 4th ed. McGraw-Hill, 2002. 1408 pp.

[87]   G. Tchobanoglous and E. Shroeder. *Water Quality: Characteristics, Modeling, Modification*. Addison-Wesley, 1985. 768 pp.

[88]   L. N. Trefethen and D. Bau. *Numerical Linear Algebra*. Society of Industrial and Applied Mathematics, 1997. 361 pp.

[89]   J. M. Tremblay, J. W. Loder, F. E. Werner, C. E. Naimie, F. H. Page, and M. M. Sinclair. "Drift of Sea Scallop Larvae *placopecten magellanicus* on Georges Bank: A Model Study of the Roles of Mean Advection, Larval Behavior and Larval Origin." *Deep Sea Research*, Pt. 2: *Topical Studies in Oceanography*, Vol. 41, No. 1 (1994): 7–49.

[90]   J. C. J. M. van den Bergh, J. Hoekstra, R. Imeson, P. A. L. D. Nunes, and A. T. de Blaeij. *Bioeconomic Modeling and Valuation of Exploited Marine Ecosystems*. Springer, 2006. 263 pp.
       — W. Ver Eecke. *See* W. Ver Eecke [22].

[91]   C. J. Walters and S. J. D. Martell. *Fisheries Ecology and Management*. Princeton University Press, 2004. 399 pp.

[92]   F. E. Werner, B. R. MacKenzie, R. I. Perry, R. G. Lough, C. E. Naimie, B. O. Blanton, and J. A. Quinlan. "Larval Trophodynamics, Turbulence, and Drift on Georges Bank: A Sensitivity Analysis of Cod and Haddock. *Scientia Marina*, Vol. 65, No. S1 (2001): 99–115.

[93]   F. E. Werner, R. I. Perry, R. G. Lough, and C. E. Naimie. "Trophodynamic and Advective Influences on Georges Bank Larval Cod and Haddock. *Deep Sea Research*, Pt. 2, Vol. 43, Nos. 7/8 (1996): 1793–1822.

[94]   F. Werner, J. Quinlan, G. Lough, and D. Lynch. "Spatially-Explicit Individual Based Modeling of Marine Populations: A Review of the Advances in the 1990's. *Sarsia*, Vol. 86, No. 6 (2001): 411–421.

[95]   (UN) World Water Assessment Programme/Unesco/Berghann Books. *Water: A Shared Responsibility*. (UN World Water Development Report, 2) United Nations Publications, 2006. 584 pp.

[96]   R. V. Thomann and J. A. Mueller, *Principles of Surface Water Quality Modeling and Control*. Harper Collins, 1987, 644 pp.

[97]   S. C. Chapra, *Surface Water-Quality Modeling*. McGraw-Hill, 1977. 844 pp.

INDEX

Note: entries followed by a lower-case f or t refer to figures or tables.